Advances in MATHEMATICAL ECONOMICS

Aims and Scope. The project is to publish *Advances in Mathematical Economics* once a year under the auspices of the Research Center for Mathematical Economics. It is designed to bring together those mathematicians who are seriously interested in obtaining new challenging stimuli from economic theories and those economists who are seeking effective mathematical tools for their research.

The scope of *Advances in Mathematical Economics* includes, but is not limited to, the following fields:

- Economic theories in various fields based on rigorous mathematical reasoning.
- Mathematical methods (e.g., analysis, algebra, geometry, probability) motivated by economic theories.
- Mathematical results of potential relevance to economic theory.
- Historical study of mathematical economics.

Authors are asked to develop their original results as fully as possible and also to give a clear-cut expository overview of the problem under discussion. Consequently, we will also invite articles which might be considered too long for publication in journals.

S. Kusuoka, T. Maruyama (Eds.)

Advances in Mathematical Economics

Volume 16

Springer

Shigeo Kusuoka
Professor
Graduate School of Mathematical Sciences
The University of Tokyo
3-8-1 Komaba, Meguro-ku
Tokyo 153-8914, Japan

Toru Maruyama
Professor
Department of Economics
Keio University
2-15-45 Mita, Minato-ku
Tokyo 108-8345, Japan

ISSN 1866-2226 1866-2234 (electronic)
ISBN 978-4-431-54113-4 978-4-431-54114-1 (eBook)
DOI 10.1007/978-4-431-54114-1
Springer Tokyo Heidelberg New York Dordrecht London

Printed on acid-free paper

Springer is part of Springer Science+Business Media (www.springer.com)

Table of Contents

Adv. Math. Econ. 16, 1–38 (2012)

Advances in
MATHEMATICAL
ECONOMICS

Some applications of Birkhoff-Kingman ergodic theorem

Charles Castaing[1] **and Marc Lavie**[2]

[1] Départment de Mathematiques, Case courrier 051, Université Montpellier II, 34095 Montpellier Cedex 5, France
(e-mail: charles.castaing@gmail.com)

[2] Laboratoire de Mathématiques appliquées, Université de Pau et des Pays de L' Adour, BP 1155, 64013, Pau cedex France
(e-mail: marc.lavie@univ-pau.fr)

Received: June 30, 2011
Revised: September 27, 2011

JEL classification: C01, C02

Mathematics Subject Classification (2010): 28B20

Abstract. We present various convergence results for multivalued ergodic theorems in Bochner-Gelfand-Pettis integration.

Key words: Conditional expectation, epiconvergence, ergodic, Bochner-Gelfand-Pettis integration, Birkhoff-Kingman ergodic theorem, Mosco convergence, multivalued convergence, slice convergence

1. Introduction

Classical ergodic theorems for real valued random variables have been recently extended into the context of epiconvergence in [7, 17, 24, 25, 34]. Using Abid result [1] on the a.s. convergence of subadditive superstationary process, Krupa [27] and Schurger [32] treated the Ergodic theorems for subadditive superstationary families of convex compact random sets. Ghoussoub and Steele [6] treated the a.s. norm convergence for subadditive process in an order complete Banach lattice extending the Kingman's theorem for real

S. Kusuoka and T. Maruyama (eds.), *Advances in Mathematical Economics Volume 16*, DOI: 10.1007/978-4-431-54114-1_1,

valued subadditive process. In this paper we present various applications of the Birkhoff-Kingman ergodic theorem. The paper is organized as follows. In Sects. 3–4 we state and summarize for references some results on the conditional expectation of closed convex valued integrable multifunctions in separable Banach spaces and in their dual spaces. Main results are given in Sects. 5–8. For the sake of completeness we provide an epiconvergence result for parametric ergodic theorem in Sect. 5 that is a starting point of this study. In Sect. 6 we treat the Mosco convergence for convex weakly compact valued ergodic theorem in Bochner integration and the weak star Kuratowski convergence for convex weakly star compact valued ergodic theorem in Gelfand integration and also a scalar convergence result for convex weakly compact valued ergodic theorem in Pettis integration. An unusual convergence for superadditive integrable process in Banach lattice is given in Sect. 7 using the integrable selection theorem for the sequential weak upper limit of a sequence of measurable closed convex valued random sets. Some relationships with economic problems are also discussed. In Sect. 8 we present a convergence theorem for convex weakly compact valued superadditive process in Bochner integration via Komlós techniques.

2. Notations and preliminaries

Throughout this paper $(\Omega, \mathcal{F}, P)$ is a complete probability space, $(\mathcal{F}_n)_{n\in\mathbf{N}}$ is an increasing sequence of sub-σ-algebras of $\mathcal{F}$ such that $\mathcal{F}$ is the σ-algebra generated by $\cup_{n\in\mathbf{N}}\mathcal{F}_n$, E is a separable Banach space and E^* is its topological dual. Let $\overline{B}_E$ (resp. $\overline{B}_{E^*}$) be the closed unit ball of E (resp. E^*) and 2^E the collection of all subsets of E. Let $cc(E)$ (resp. $cwk(E)$) (resp. $\mathcal{L}wk(E)$) (resp. $\mathcal{R}wk(E)$) be the set of nonempty closed closed convex (resp. convex weakly compact) (resp. closed convex *weakly locally compact subsets of E which contain no lines*) (resp. *ball-weakly compact* closed convex) subsets of E, here a closed convex subset in E is *ball-weakly compact* if its intersection with any closed ball in E is weakly compact. For $A \in cc(E)$, the distance and the support function associated with A are defined respectively by

$$d(x, A) = \inf\{\|x - y\| \ : y \in A\}, (x \in E)$$

$$\delta^*(x^*, A) = \sup\{\langle x^*, y\rangle \ : y \in A\}, (x^* \in E^*).$$

We also define

$$|A| = \sup\{||x|| \ : x \in A\}.$$

A sequence $(K_n)_{n\in\mathbf{N}}$ in $cwk(E)$ *scalarly converges* to $K_\infty \in cwk(E)$ if $\lim_{n\to\infty} \delta^*(x^*, K_n) = \delta^*(x^*, K_\infty), \forall x^* \in E^*$. Let B be a closed bounded convex subset of E and let C be the closed convex subset of E. Then the gap [4] $D(B, C)$ between B and C is denoted by

$$D(B, C) = \inf\{||x - y|| : x \in B, y \in C\}.$$

By Hahn Banach theorem we know that

$$D(B, C) = \sup_{x^* \in \overline{B}_{E^*}} \{-\delta^*(x^*, C) - \delta^*(-x^*, B)\}.$$

Given a sub-σ-algebra $\mathcal{B}$ in Ω, a mapping $X : \Omega \to 2^E$ is $\mathcal{B}$-measurable if for every open set U in E the set

$$X^- U := \{\omega \in \Omega : X(\omega) \cap U \neq \emptyset\}$$

is a member of $\mathcal{B}$. A function $f : \Omega \to E$ is a $\mathcal{B}$-measurable selection of X if $f(\omega) \in X(\omega)$ for all $\omega \in \Omega$. A Castaing representation of X is a sequence $(f_n)_{n \in \mathbf{N}}$ of $\mathcal{B}$-measurable selections of X such that

$$X(\omega) = cl\{f_n(\omega), n \in \mathbf{N}\} \quad \forall \omega \in \Omega$$

where the closure is taken with respect to the topology of associated with the norm in E. It is known that a nonempty closed-valued multifunction $X : \Omega \to c(E)$ is $\mathcal{B}$-measurable iff it admits a Castaing representation. If $\mathcal{B}$ is complete, the $\mathcal{B}$-measurability is equivalent to the measurability in the sense of graph, namely the graph of X is a member of $\mathcal{B} \otimes \mathcal{B}(E)$, here $\mathcal{B}(E)$ denotes the Borel tribe on E. A $cc(E)$-valued $\mathcal{B}$-measurable $X : \Omega \to cc(E)$ is integrable if the set $S^1_X(\mathcal{B})$ of all $\mathcal{B}$-measurable and integrable selections of X is nonempty. We denote by $L^1_E(\mathcal{B})$ the space of E-valued $\mathcal{B}$-measurable and Bochner-integrable functions defined on Ω and $\mathcal{L}^1_{cwk(E)}(\mathcal{B})$ the space of all $\mathcal{B}$-measurable multifunctions $X : \Omega \to cwk(E)$ such that $|X| \in L^1_{\mathbf{R}}(\mathcal{B})$. We refer to [16] for the theory of Measurable Multifunctions and Convex Analysis, and to [18, 29] for Real Analysis and Probability.

3. Multivalued conditional expectation theorem

Given a sub-σ-algebra $\mathcal{B}$ of $\mathcal{F}$ and an integrable $\mathcal{F}$-measurable $cc(E)$-valued multifunction $X : \Omega \rightrightarrows E$, Hiai and Umegaki [20] showed the existence of a $\mathcal{B}$-measurable $cc(E)$-valued integrable multifunction denoted by $E^{\mathcal{B}} X$ such that

$$\mathcal{S}^1_{E^{\mathcal{B}} X}(\mathcal{B}) = cl\{E^{\mathcal{B}} f : f \in \mathcal{S}^1_X(\mathcal{F})\}$$

the closure being taken in $L^1_E(\Omega, \mathcal{A}, P)$; $E^{\mathcal{B}} X$ is the multivalued conditional expectation of X relative to $\mathcal{B}$. If $X \in \mathcal{L}^1_{cwk(E)}(\mathcal{F})$ and the strong dual E^*_b is separable, then $E^{\mathcal{B}}X \in \mathcal{L}^1_{cwk(E)}(\mathcal{B})$ with $\mathcal{S}^1_{E^{\mathcal{B}}X}(\mathcal{B}) = \{E^{\mathcal{B}}f : f \in \mathcal{S}^1_X(\mathcal{F})\}$. A unified approach for general conditional expectation of $cc(E)$-valued integrable multifunctions is given in [33] allowing to recover both the $cc(E)$-valued conditional expectation of $cc(E)$-valued integrable multifunctions in the sense of [20] and the $cwk(E)$-valued conditional expectation of $cwk(E)$-valued integrably bounded multifunctions given in [5]. For more information on multivalued conditional expectation and related subjects we refer to [2, 9, 16, 20, 24, 33]. In the context of this paper we summarize a specific version of conditional expectation in a separable Banach space.

Proposition 3.1. *Assume that the strong dual E^*_b is separable. Let $\mathcal{B}$ be a sub-σ-algebra of $\mathcal{F}$ and an integrable $\mathcal{F}$-measurable $cc(E)$-valued multifunction $X : \Omega \rightrightarrows E$. Assume further there is a $\mathcal{F}$-measurable ball-weakly compact $cc(E)$-valued multifunction $K : \Omega \rightrightarrows E$ such that $X(\omega) \subset K(\omega)$ for all $\omega \in \Omega$. Then there is a unique (for the equality a.s.) $\mathcal{B}$-measurable $cc(E)$-valued multifunction Y satisfying the property*

$$(^*)\quad \forall v \in L^\infty_{E^*}(\mathcal{B}), \int_\Omega \delta^*(v(\omega), Y(\omega))dP(\omega) = \int_\Omega \delta^*(v(\omega), X(\omega))dP(\omega).$$

$E^{\mathcal{B}}X := Y$ is the conditional expectation of X.

Proof. The proof is an adaptation of the one of Theorem VIII.35 in [16]. Let u_0 be an integrable selection of X. For every $n \in \mathbf{N}$, let

$$X_n(\omega) = X(\omega) \cap (u_0(\omega) + n\overline{B}_E) \quad \forall n \in \mathbf{N} \quad \forall \omega \in \Omega.$$

As $X(\omega) \subset K(\omega)$ for all $\omega \in \Omega$, we get

$$X_n(\omega)=X(\omega)\cap(u_0(\omega)+n\overline{B}_E)\subset K(\omega)\cap(u_0(\omega)+n\overline{B}_E) \quad \forall n \in \mathbf{N} \quad \forall \omega \in \Omega.$$

As $K(\omega)$ is ball-weakly compact, it is immediate that $X_n \in \mathcal{L}^1_{cwk(E)}(\mathcal{F})$. so that, by virtue of ([5] or ([33], Remarks of Theorem 3), the conditional expectation $E^{\mathcal{B}}X_n \in \mathcal{L}^1_{cwk(E)}(\mathcal{B})$. It follows that

$$(^{**})\qquad \int_\Omega \delta^*(v(\omega), E^{\mathcal{B}}X_n(\omega))P(d\omega) = \int_\Omega \delta^*(v(\omega), X_n(\omega))P(d\omega)$$

$\forall n \in \mathbf{N}, \forall v \in L^\infty_{E^*}(\mathcal{B})$. Now let

$$Y(\omega) = cl(\cup_{n\in\mathbf{N}} E^{\mathcal{B}}X_n(\omega)) \quad \forall \omega \in \Omega.$$

Then Y is $\mathcal{B}$-measurable and a.s. convex. Now the required property (*) follows from (**) and the monotone convergence theorem. Indeed

$$\forall n \in \mathbf{N}, \forall v \in L^{\infty}_{E^*}(\mathcal{B}), \langle u_0, v\rangle \leq \delta^*(v, X_n) \uparrow \delta^*(v, X)$$
$$\langle v, E^{\mathcal{B}} u_0\rangle \leq \delta^*(v, E^{\mathcal{B}} X_n) \uparrow \delta^*(v, Y).$$

Now the uniqueness follows exactly as in the proof of Theorem VIII.35 in [16], via the measurable projection theorem ([16], Theorem III.32). ■

4. Conditional expectation in a dual space

Let $(\Omega, \mathcal{F}, P)$ be a complete probability space, $(\mathcal{F}_n)_{n\in\mathbf{N}}$ an increasing sequence of sub σ-algebras of $\mathcal{F}$ such that $\mathcal{F}$ is the σ-algebra generated by $\cup_{n\geq 1}^{\infty}\mathcal{F}_n$. Let E be a separable Banach space, $D_1 = (x_p)_{p\in\mathbf{N}}$ is a dense sequence in the closed unit ball of E, E^* is the topological dual of E, $\overline{B}_E$ (resp. $\overline{B}_{E^*}$) is the closed unit ball of E (resp. E^*). We denote by E^*_s (resp. E^*_c) (resp. E^*_b) the topological dual E^* endowed with the topology $\sigma(E^*, E)$ of pointwise convergence, alias w^* topology (resp. the topology τ_c of compact convergence) (resp. the topology s^* associated with the dual norm $||.||_{E^*_b}$), and by $E^*_{m^*}$ the topological dual E^* endowed with the topology $m^* = \sigma(E^*, H)$, where H is the linear space of E generated by D, that is the Hausdorff locally convex topology defined by the sequence of semi-norms

$$p_k(x^*) = \max\{|\langle x^*, x_p\rangle| : p \leq k\}, \ \ x^* \in E^*, k \geq 1.$$

Recall that the topology m^* is metrizable by the metric

$$d_{E^*_{m^*}}(x^*_1, x^*_2) := \sum_{p=1}^{p=\infty} \frac{1}{2^p}|\langle x_p, x^*_1\rangle - \langle x_p, x^*_2\rangle|, \quad x^*_1, x^*_2 \in E^*.$$

Further we have

$$d_{E^*_{m^*}}(x^*, y^*) \leq ||x^* - y^*||_{E^*_b}, \quad \forall x^*, y^* \in E^* \times E^*.$$

We assume from now that $d_{E^*_{m^*}}$ is held fixed. Further, we have $m^* \subset w^* \subset \tau_c \subset s^*$. When E is infinite dimensional these inclusions are strict. On the other hand, the restrictions of m^*, w^*, to any bounded subset of E^* coincide and the Borel tribes $\mathcal{B}(E^*_s)$, $\mathcal{B}(E^*_c)$ and $\mathcal{B}(E^*_{m^*})$ associated with E^*_s, E^*_c and $E^*_{m^*}$ are equal. Noting that E^* is the countable union of closed balls, we deduce that the space E^*_s is Suslin, as well as the metrizable topological space $E^*_{m^*}$. A $2^{E^*_s}$-valued multifunction (alias mapping for short) $X : \Omega \Rightarrow E^*_s$ is

$\mathcal{F}$-measurable if its graph belongs to $\mathcal{F} \otimes \mathcal{B}(E_s^*)$. Given a $\mathcal{F}$-measurable mapping $X : \Omega \rightrightarrows E_s^*$ and a Borel set $G \in \mathcal{B}(E_s^*)$, the set

$$X^- G = \{\omega \in \Omega : X(\omega) \cap G \neq \emptyset\}$$

is $\mathcal{F}$-measurable, that is $X^- G \in \mathcal{F}$. In view of the completeness hypothesis on the probability space, this is a consequence of the Projection Theorem (see e.g. Theorem III.23 of [16]) and the equality

$$X^- G = \mathrm{proj}_\Omega \{Gr(X) \cap (\Omega \times G)\}.$$

Further if $u : \Omega \to E_s^*$ is a scalarly $\mathcal{F}$-measurable mapping, that is, for every $x \in E$, the scalar function $\omega \mapsto \langle x, u(\omega)\rangle$ is $\mathcal{F}$-measurable, then the function $f : (\omega, x^*) \mapsto ||x^* - u(\omega)||_{E_b^*}$ is $\mathcal{F} \otimes \mathcal{B}(E_s^*)$-measurable, and for every fixed $\omega \in \Omega$, $f(\omega, .)$ is lower semicontinuous on E_s^*, shortly, f is a normal integrand, indeed, we have

$$||x^* - u(\omega)||_{E_b^*} = \sup_{j \in \mathbf{N}} \langle e_j, x^* - u(\omega)\rangle$$

here $D_1 = (e_j)_{j \geq 1}$ is a dense sequence in the closed unit ball of E. As each function $(\omega, x^*) \mapsto \langle e_j, x^* - u(\omega)\rangle$ is $\mathcal{F} \otimes \mathcal{B}(E_s^*)$-measurable and continuous on E_s^* for each $\omega \in \Omega$, it follows that f is a normal integrand. Consequently, the graph of u belongs to $\mathcal{F} \otimes \mathcal{B}(E_s^*)$. Besides these facts, let us mention that the function $d_{E_b^*}(x^*, y^*) = ||x^* - y^*||_{E_b^*}$ is lower semicontinuous on $E_s^* \times E_s^*$, being the supremum of w^*-continuous functions. If X is a $\mathcal{F}$-measurable mapping, the distance function $\omega \mapsto d_{E_b^*}(x^*, X(\omega))$ is $\mathcal{F}$-measurable, by using the lower semicontinuity of the function $d_{E_b^*}(x^*, .)$ on E_s^* and measurable projection theorem ([16], Theorem III.23) and recalling that E_s^* is a Suslin space. A mapping $u : \Omega \rightrightarrows E_s^*$ is said to be scalarly integrable, if, for every $x \in E$, the scalar function $\omega \mapsto \langle x, u(\omega)\rangle$ is $\mathcal{F}$-measurable and integrable. We denote by $L^1_{E^*}[E](\mathcal{F})$ the subspace of all $\mathcal{F}$-measurable mappings u such that the function $|u| : \omega \mapsto ||u(\omega)||_{E_b^*}$ is integrable. The measurability of $|u|$ follows easily from the above considerations. By $cwk(E_s^*)$ we denote the set of all nonempty convex $\sigma(E^*, E)$-compact subsets of E_s^*. A $cwk(E_s^*)$-valued mapping $X : \Omega \rightrightarrows E_s^*$ is scalarly $\mathcal{F}$-measurable if the function $\omega \to \delta^*(x, X(\omega))$ is $\mathcal{F}$-measurable for every $x \in E$. Let us recall that any scalarly $\mathcal{F}$-measurable $cwk(E_s^*)$-valued mapping is $\mathcal{F}$-measurable. Indeed, let $(e_k)_{k \in \mathbf{N}}$ be a sequence in E which separates the points of E^*, then we have $x \in X(\omega)$ iff, $\langle e_k, x\rangle \leq \delta^*(e_k, X(\omega))$ for all $k \in \mathbf{N}$. By $\mathcal{L}^1_{cwk(E_s^*)}(\Omega, \mathcal{F}, P)$ (shortly $\mathcal{L}^1_{cwk(E_s^*)}(\mathcal{F})$) we denote the of all scalarly integrable $cwk(E)$-valued multifunctions X such that the function $|X| : \omega \to |X(\omega)|$ is integrable, here $|X(\omega)| := \sup_{y^* \in X(\omega)} ||y^*||_{E_b^*}$, by the above consideration, it is easy to see that $|X|$ is $\mathcal{F}$-measurable. For the convenience of the reader we recall and summarize the existence and uniqueness of the conditional expectation in $\mathcal{L}^1_{cwk(E^*_s)}(\mathcal{F})$. See ([33], Theorem 3).

Theorem 4.1. *Given a $\Gamma \in \mathcal{L}^1_{cwk(E^*_s)}(\mathcal{F})$ and a sub-σ-algebra $\mathcal{B}$ of $\mathcal{F}$, there exists a unique (for equality a.s.) mapping $\Sigma := E^{\mathcal{B}}\Gamma \in \mathcal{L}^1_{cwk(E^*_s)}(\mathcal{B})$, that is the conditional expectation of Γ with respect to $\mathcal{B}$, which enjoys the following properties:*

a) $\int_\Omega \delta^*(v, \Sigma)dP = \int_\Omega \delta^*(v, \Gamma)dP$ *for all* $v \in L^\infty_E(\mathcal{B})$.
b) $\Sigma \subset E^{\mathcal{B}}|\Gamma|\overline{B}_{E^*}$ *a.s.*
c) $\mathcal{S}^1_\Sigma(\mathcal{B})$ *is sequentially* $\sigma(L^1_{E^*}[E](\mathcal{B}), L^\infty_E(\mathcal{B}))$ *compact (here* $\mathcal{S}^1_\Sigma(\mathcal{B})$ *denotes the set of all* $L^1_{E^*}[E](\mathcal{B})$ *selections of* Σ*) and satisfies the inclusion*

$$E^{\mathcal{B}}\mathcal{S}^1_\Gamma(\mathcal{F}) \subset \mathcal{S}^1_\Sigma(\mathcal{B}).$$

d) Furthermore one has

$$\delta^*(v, E^{\mathcal{B}}\mathcal{S}^1_\Gamma(\mathcal{F})) = \delta^*(v, \mathcal{S}^1_\Sigma(\mathcal{B}))$$

for all $v \in L^\infty_E(\mathcal{B})$.
e) $E^{\mathcal{B}}$ *is increasing:* $\Gamma_1 \subset \Gamma_2$ *a.s. implies* $E^{\mathcal{B}}\Gamma_1 \subset E^{\mathcal{B}}\Gamma_2$ *a.s.*

This result involves the existence of conditional expectation for $\sigma(E^*, E)$ closed convex integrable mapping in E^*, namely

Theorem 4.2. *Given a $\mathcal{F}$-measurable $\sigma(E^*, E)$ closed convex mapping Γ in E^* which admits a integrable selection $u_0 \in L^1_{E^*}[E](\mathcal{F})$ and a sub-σ-algebra $\mathcal{B}$ of $\mathcal{F}$. For every $n \in \mathbf{N}$ and for every $\omega \in \Omega$ let*

$$\Gamma_n(\omega) = \Gamma(\omega) \cap (u_0(\omega) + n\overline{B}_{E^*}).$$

$$\Sigma(\omega) = \sigma(E^*, E) - cl[\cup E^{\mathcal{B}}\Gamma_n(\omega)].$$

Then $\Sigma(\omega)$ is a.s. convex and is a $\mathcal{B}$-measurable $\sigma(E^*, E)$ closed convex mapping which satisfies the properties:

a) $\int_\Omega \delta^*(v, \Sigma)dP = \int_\Omega \delta^*(v, \Gamma)dP$ *for all* $v \in L^\infty_E(\mathcal{B})$.
b) $\Sigma := E^{\mathcal{B}}\Gamma$ *is the unique for equality a.s.* $\mathcal{B}$*-measurable* $\sigma(E^*, E)$ *closed convex mapping with property a).*
c) $E^{\mathcal{B}}\Gamma$ *is increasing:* $\Gamma_1 \subset \Gamma_2$ *a.s. implies* $E^{\mathcal{B}}\Gamma_1 \subset E^{\mathcal{B}}\Gamma_2$ *a.s.*
By definition, $\Sigma := E^{\mathcal{B}}\Gamma$ *is the conditional expectation of* Γ.

Proof. Follows the same line of the proof of Theorem VIII-35 in [16] and is omitted. ■

Before going further we state and summarize an epiconvergence result [6] which is a starting point of our study.

5. An epiconvergence result

The following is a special version of an epiconvergence result in [6]. We summarize and give the details of proof because it has an own interest and leads to the convergence of multivalued Birkhoff's ergodic theorem.

Proposition 5.1. *Let E be a separable Banach space, T a measurable transformation of Ω preserving P, $\mathcal{I}$ the σ-algebra of invariant sets and $f : \Omega \times E^*_s \to \overline{\mathbf{R}}$ be a $\mathcal{F} \times \mathcal{B}(E^*_s)$-measurable normal convex integrand such that $f^+(\omega, 0) = 0$ for all $\omega \in \Omega$ and let f^* the polar of f. Assume that*

$$\int_\Omega f^*(\omega, u(\omega)) dP(\omega) < \infty \quad \forall u \in L^\infty_{\overline{B}_E}(\Omega, \mathcal{I}, P).$$

The following epiconvergence result holds, for any $u \in L^\infty_{\overline{B}_E}(\Omega, \mathcal{I}, P)$, one has:

$$(^*) \quad \sup_{k \in \mathbf{N}} \limsup_{n\to\infty} \inf_{y \in \overline{B}_E} [\frac{1}{n}\sum_{j=0}^{n-1} f^*(T^j\omega, y) + k||u(\omega) - y||]$$
$$\leq g^*(\omega, u(\omega)) = E^{\mathcal{I}} f^*(\omega, u(\omega)) \quad a.s.$$

where

$$g^*(\omega, x) = \delta^*((x, -1), E^{\mathcal{I}} epi f(\omega, .)) \quad \forall(\omega, x) \in \Omega \times E.$$

and

$$(^{**}) \quad \sup_{k \in \mathbf{N}} \liminf_{n\to\infty} \inf_{y \in \overline{B}_E} [\frac{1}{n}\sum_{j=0}^{n-1} f^*(T^j\omega, y) + k||u(\omega) - y||]$$
$$\geq g^*(\omega, u(\omega)) = E^{\mathcal{I}} f^*(\omega, u(\omega)) \quad a.s.$$

Proof. Step 1 Here we use some arguments developed in the proof of Proposition 5.5 in [7]. Let

$$\Gamma(\omega) := epi f(\omega, .) \quad \forall \omega \in \Omega.$$

Then Γ is a closed convex integrable mapping in $E^*_s \times \mathbf{R}$ and admits an integrable selection, namely

$$(0, 0) \in \Gamma(\omega) \subset E^*_s \times \mathbf{R} \quad \forall \omega \in \Omega.$$

Let $E^{\mathcal{B}}\Gamma$ be the conditional expectation of Γ whose existence is given by Proposition 4.2. Then the conditional expectation g^* of the normal integrand f^* is a $\mathcal{I} \otimes \mathcal{B}(E)$-measurable normal integrand satisfying

$$\int_A f^*(\omega, v(\omega)) dP(\omega) = \int_A g^*(\omega, v(\omega)) dP(\omega)$$

for all $v \in L^\infty_E(\Omega, \mathcal{I}, P)$, and for all $A \in \mathcal{I}$ with

$$f^*(\omega, v(\omega)) := \delta^*((v(\omega), -1), epif(\omega, .)) \quad \forall \omega \in \Omega$$
$$g^*(\omega, v(\omega)) := \delta^*((v(\omega), -1), E^{\mathcal{I}} epif(\omega, .)) \quad \forall \omega \in \Omega.$$

Hence $g^*(\omega, v(\omega)) = E^{\mathcal{I}} f^*(\omega, v(\omega))$ is positive and integrable for all $v \in L^\infty_{\overline{B}_E}(\Omega, \mathcal{I}, P)$. Let us set

$$\Phi(\omega, x) := g^*(\omega, x) \quad \forall (\omega, x) \in \Omega \times \overline{B}_E.$$

and

$$\Phi^k(\omega, x) = \inf_{y \in \overline{B}_E} [\Phi(\omega, y) + k||x - y||] \quad \forall (\omega, x) \in \Omega \times \overline{B}_E.$$

Then

$$\sup_{k \in \mathbf{N}} \Phi^k(\omega, x) = \Phi(\omega, x) \quad \forall (\omega, x) \in \Omega \times \overline{B}_E.$$

Let $u \in L^\infty_{\overline{B}_E}(\Omega, \mathcal{I}, P)$. Let $p \in \mathbf{N}$. Since E is separable Banach space, applying measurable selection theorem ([16], Theorem III-22) yields a $(\mathcal{I}, \mathcal{B}(E))$-measurable mapping $v_{k,p,u} : \Omega \to \overline{B}_E$ such that

$$0 \leq \Phi(\omega, v_{k,p,u}(\omega)) + k||u(\omega) - v_{k,p,u}(\omega)|| \leq \Phi^k(\omega, u(\omega)) + \frac{1}{p} \quad \forall \omega \in \Omega.$$

From the classical Birkhoff ergodic theorem (see e.g. [34]) we provide a negligible set $N_{k,p,u}$ such that for $\omega \notin N_{k,p,u}$

$$\lim_{n \to \infty} \frac{1}{n} \sum_{j=0}^{n-1} f^*(T^j\omega, v_{k,p,u}(\omega)) = E^{\mathcal{I}}[f^*(\omega, v_{k,p,u}(\omega))]$$
$$= \Phi(\omega, v_{k,p,u}(\omega)) \quad a.s.$$

Whence we provide a negligible set $N_{k,p,u}$ such that

$$\limsup_{n \to \infty} \inf_{y \in \overline{B}_E} [\frac{1}{n} \sum_{j=0}^{n-1} f^*(T^j\omega, y) + k||u(\omega) - y||_E]$$
$$\leq \limsup_{n \to \infty} [\frac{1}{n} \sum_{j=0}^{n-1} f^*(T^j\omega, v_{k,p,u}(\omega)) + k||u(\omega) - v_{k,p,u}(\omega)||_E]$$
$$= \Phi(\omega, v_{k,p,u}(\omega)) + k||u(\omega) - v_{k,p,u}(\omega)||_E \leq \Phi^k(\omega, u(\omega)) + \frac{1}{p} \quad \forall \omega \notin N_{k,p,u}.$$

Taking the supremum on $k \in \mathbf{N}$ in this inequality yields a negligible set $N_u := \cup_{k\in\mathbf{N}, p\in\mathbf{N}} N_{k,p,u}$ such that for $\omega \notin N_u$

$$\sup_{k\in\mathbf{N}} \limsup_{n\to\infty} \inf_{y\in\overline{B}_E} [\frac{1}{n}\sum_{j=0}^{n-1} f^*(T^j\omega, y) + k||u(\omega) - y||_E]$$

$$\leq \sup_{k\in\mathbf{N}} \Phi^k(\omega, u(\omega)) = \Phi(\omega, u(\omega)) = g^*(\omega, u(\omega)).$$

Step 2 For $k \in \mathbf{N}$, $n \in \mathbf{N}$ and for $(\omega, x) \in \Omega \times \overline{B}_E$, we set

$$\Psi(\omega, x) := f^*(\omega, x)$$

$$\Psi^k(\omega, x) = \inf_{y\in\overline{B}_E} [f^*(\omega, y) + k||x - y||_E].$$

$$\Psi_n(\omega, x) = \frac{1}{n}\sum_{j=0}^{n-1} f^*(T^j\omega, x) = \frac{1}{n}\sum_{j=0}^{n-1} \Psi(T^j\omega, x).$$

$$\Psi_n^k(\omega, x) = \inf_{y\in\overline{B}_E} [\Psi_n(\omega, y) + k||x - y||_E].$$

Then the following hold

(5.1.1) $|\Psi^k(\omega, x) - \Psi^k(\omega, y)| \leq k||x - y||_E \quad \forall(\omega, x, y) \in \Omega \times \overline{B}_E \times \overline{B}_E.$

(5.1.2) $$0 \leq \Psi^k(\omega, x) \leq \Psi(\omega, x) \quad \forall(\omega, x) \in \Omega \times \overline{B}_E.$$

(5.1.3) $$\sup_{k\in\mathbf{N}} \Psi^k(\omega, x) = \Psi(\omega, x) \quad \forall(\omega, x) \in \Omega \times \overline{B}_E.$$

There is a negligible set N which does not depend on $x \in \overline{B}_E$ such that

(5.1.4) $$E^{\mathcal{I}}\Psi(\omega, x) = \begin{cases} \sup_{k\in\mathbf{N}} E^{\mathcal{I}}\Psi^k(\omega, x) & \text{if} \quad \omega \in \Omega \setminus N \times \overline{B}_E \\ 0 \quad \text{if} \quad (\omega, x) \in N \times \overline{B}_E \end{cases}$$

where $E^{\mathcal{I}}\Psi^k$ and $E^{\mathcal{I}}\Psi$ denote the conditional expectation relative to $\mathcal{I}$ of Ψ^k and Ψ respectively. Further we have

(5.1.5) $$\Psi_n^k(\omega, x) \geq \frac{1}{n}\sum_{j=0}^{n-1} \Psi^k(T^j\omega, x) \quad \forall(\omega, x) \in \Omega \times \overline{B}_E.$$

By virtue of classical Birkhoff ergodic theorem (see e.g. Lemma 5 in [34]) it follows that

$$(5.1.6)\qquad \lim_{n\to\infty}\frac{1}{n}\sum_{j=0}^{n-1}\Psi^k(T^j\omega,u(\omega)) = E^{\mathcal{I}}[\Psi^k(\omega,u(\omega))]\quad a.s.$$

(5.1.5) -(5.1.6) yield a negligible set $N_{k,u}$ such that

$$\begin{aligned}\liminf_{n\to\infty}\inf_{y\in\overline{B}_E}&[\frac{1}{n}\sum_{j=0}^{n-1}\Psi(T^j\omega,y)+k||u(\omega)-y||_E]\\ &\geq \liminf_{n\to\infty}[\frac{1}{n}\sum_{j=0}^{n-1}\Psi^k(T^j\omega,u(\omega))]\\ &= E^{\mathcal{I}}[\Psi^k(\omega,u(\omega))]\quad \forall\omega\notin N_{k,u}.\end{aligned}$$

Using (5.1.4) and the preceding convergences, we produce a negligible set $N_u = \cup_{k\in\mathbf{N}}N_{k,u}\cup N$ such that

$$\begin{aligned}\sup_{k\in\mathbf{N}}\liminf_{n\to\infty}\inf_{y\in\overline{B}_E}&[\frac{1}{n}\sum_{j=0}^{n-1}\Psi(T^j\omega,y)+k||u(\omega)-y||_E]\\ &\geq E^{\mathcal{I}}\Psi(\omega,u(\omega)) = g^*(\omega,u(\omega))\quad\forall\omega\notin N_u.\end{aligned}$$

■

For more on epiconvergence results in ergodicity one may consult [6, 7, 17, 24, 34].

6. Birkhoff ergodic theorem for Bochner-Gelfand-Pettis integrable multifunctions

We need to recall some classical results on the slice convergence. For more information on the slice topology one may consult [3, 4].

Lemma 6.1. *Assume that E^* is strongly separable. Let D_1^* be a dense sequence in the closed unit ball $\overline{B}_{E^*}$ of E^*. Then for all bounded closed convex subsets B and C in X, the following holds:*

$$\begin{aligned}D(B,C) :&= \sup_{x^*\in\overline{B}_{E^*}}\{-\delta^*(x^*,C)-\delta^*(-x^*,B)\}\\ &= \sup_{x^*\in D_1^*}\{-\delta^*(x^*,C)-\delta^*(-x^*,B)\}.\end{aligned}$$

Proof. Equality

$$D(B,C) = \sup_{x^*\in \overline{B}_{E^*}} \{-\delta^*(x^*,C) - \delta^*(-x^*,B)\}$$

follows from Hahn-Banach theorem, while the second equality

$$\sup_{x^*\in \overline{B}_{E^*}} \{-\delta^*(x^*,C) - \delta^*(-x^*,B)\} = \sup_{x^*\in D_1^*} \{-\delta^*(x^*,C) - \delta^*(-x^*,B)\}$$

follows from the strong separability of E^*, noting that the function $x^* \to -\delta^*(x^*,C) - \delta^*(-x^*,B)$ is strongly continuous on E^*.

Lemma 6.2. *Assume that E is separable. Let D_1^* be a dense sequence in the closed unit ball $\overline{B}_{E^*}$ of E^* with respect to the Mackey topology. Then for all convex weakly compact subsets B and C in E, the following holds:*

$$\begin{aligned} D(B,C) :&= \sup_{x^*\in \overline{B}_{E^*}} \{-\delta^*(x^*,C) - \delta^*(-x^*,B)\} \\ &= \sup_{x^*\in D_1^*} \{-\delta^*(x^*,C) - \delta^*(-x^*,B)\}. \end{aligned}$$

Proof. Equality

$$D(B,C) = \sup_{x^*\in \overline{B}_{E^*}} \{-\delta^*(x^*,C) - \delta^*(-x^*,B)\}$$

follows from Hahn-Banach theorem, while equality

$$\sup_{x^*\in \overline{B}_{E^*}} \{-\delta^*(x^*,C) - \delta^*(-x^*,B)\} = \sup_{x^*\in D_1^*} \{-\delta^*(x^*,C) - \delta^*(-x^*,B)\}$$

follows, noting that the function $x^* \to -\delta^*(x^*,C) - \delta^*(-x^*,B)$ is continuous on E^* with respect to the Mackey topology.

We begin to treat the Mosco convergence for multivalued ergodic theorem in Bochner integration.

Theorem 6.1. *Assume that E and its strong dual E_b^* are separable. Let T be a measurable transformation of Ω preserving P, $\mathcal{I}$ the σ-algebra of invariant sets and $X : \Omega \to cwk(E)$ a $\mathcal{F}$-measurable and integrably bounded mapping, i.e. $X \in \mathcal{L}^1_{cwk(E)}(\Omega, \mathcal{F}, P)$, then the following hold:*

$$(*) \qquad \lim_{n\to\infty} \delta^*(x^*, \frac{1}{n}\sum_{j=0}^{n-1} X(T^j\omega)) = \delta^*(x^*, E^{\mathcal{I}}X(\omega))$$

for all $x^* \in \overline{B}_{E^*}$ *and almost surely* $\omega \in \Omega$, *and*

$$(^{**}) \qquad \lim_{n\to\infty} d(x, \frac{1}{n}\sum_{j=0}^{n-1} X(T^j\omega)) = d(x, E^{\mathcal{I}}X(\omega))$$

for all $x \in E$ *and almost surely* $\omega \in \Omega$, *consequently*

$$(^{***}) \qquad \lim_{n\to\infty} D(B, \frac{1}{n}\sum_{j=0}^{n-1} X(T^j\omega)) = D(B, E^{\mathcal{I}}X(\omega))$$

for all closed bounded convex subsets B of E and almost surely $\omega \in \Omega$. *Here* $E^{\mathcal{I}}X$ *is the conditional expectation of X in the sense of Hiai-Umegaki.*

Proof. Let $D_1^* = (e_k^*)_{k\in\mathbf{N}}$ be a dense sequence in the closed unit ball $\overline{B}_{E^*}$ with respect to the the norm dual topology. Note that the mapping $|X|$

$$|X|(\omega) = \sup_{x^*\in\overline{B}_{E^*}} |\delta^*(x^*, X(\omega))| \quad \forall\omega \in \Omega$$

is $\mathcal{F}$-measurable and integrable (see e.g. [16]). Next, applying the classical Birkhoff ergodic theorem to $|X|$ and each $\delta^*(e_k^*, X)$ yields almost surely

$$(6.1.1) \qquad \lim_{n\to\infty} \frac{1}{n}\sum_{j=0}^{n-1} |X|(T^j\omega) = E^{\mathcal{I}}|X|(\omega).$$

$$\lim_{n\to\infty} \delta^*(e_k^*, \frac{1}{n}\sum_{j=0}^{n-1} X(T^j\omega)) = \lim_{n\to\infty} \frac{1}{n}\sum_{j=0}^{n-1} \delta^*(e_k^*, X(T^j\omega))$$

$$(6.1.2) \qquad = \delta^*(e_k^*, E^{\mathcal{I}}X(\omega)).$$

Claim (1):

$$\liminf_{n\to\infty} D(B, \frac{1}{n}\sum_{j=0}^{n-1} X(T^j\omega)) \geq D(B, E^{\mathcal{I}}X(\omega))$$

for all closed bounded convex subsets $B \subset E$ and almost surely $\omega \in \Omega$. Indeed, for each $k \in \mathbf{N}$, there is a negligible set N_k such that (6.1.2) holds. Then $N := \cup_k N_k$ is negligible and for $\omega \in \Omega \setminus N$ we have

$$\liminf_{n\to\infty} D(B, \frac{1}{n}\sum_{j=0}^{n-1} X(T^j\omega))$$

$$= \liminf_{n\to\infty} \sup_{x\in \overline{B}_{E^*}} [-\delta^*(x^*, \frac{1}{n}\sum_{j=0}^{n-1} X(T^j\omega)) - \delta^*(-x^*, B)|$$

$$\geq \sup_{x^*\in D_1^*} \liminf_{n\to\infty}[-\delta^*(x^*, \frac{1}{n}\sum_{j=0}^{n-1} X(T^j\omega)) - \delta^*(-x^*, B)]$$

$$= \sup_{x^*\in D_1^*} [-\limsup_{n\to\infty} \delta^*(x^*, \frac{1}{n}\sum_{j=0}^{n-1} X(T^j\omega)) - \delta^*(-x^*, B)]$$

$$= \sup_{x^*\in D_1^*} [-\delta^*(x^*, E^{\mathcal{I}}X(\omega)) - \delta^*(-x^*, B)] = D(B, E^{\mathcal{I}}X(\omega)).$$

Claim (2):

$$E^{\mathcal{I}}X(\omega) \subset s\text{-}li \frac{1}{n}\sum_{j=0}^{n-1} X(T^j\omega) \quad a.s.$$

Recall that

$$S^1_{E^{\mathcal{I}}X} = cl\{E^{\mathcal{I}}f : f \in S^1_X\}$$

here the *closure* is taken in the sense of the norm in L^1_E (see also Theorem 3 in [33]). Let $f \in \mathcal{S}^1_X$. Applying Birkhoff ergodic theorem for Banach valued functions (see [26], Theorem 2.1, p.167) yields

$$\lim_{n\to\infty} \frac{1}{n}\sum_{j=0}^{n-1} f(T^j\omega) = E^{\mathcal{I}}f(\omega) \quad a.s.$$

Whence for a.s. $\omega \in \Omega$, $E^{\mathcal{I}}f(\omega)$ is the strong limit of $\frac{1}{n}\sum_{j=0}^{n-1} f(T^j\omega) \in S_n(\omega) := \frac{1}{n}\sum_{j=0}^{n-1} X(T^j\omega)$. In other words

$$E^{\mathcal{I}}f(\omega) \in s\text{-}li \frac{1}{n}\sum_{j=0}^{n-1} X(T^j\omega) \quad a.s.$$

Now if $g \in S^1_{E^{\mathcal{I}}X}$, there is a sequence (f_n) in $\mathcal{S}^1_X$ such that $E^{\mathcal{I}}f_n \to g$ a.s. so that

$$g(\omega) \in s\text{-}li \frac{1}{n}\sum_{j=0}^{n-1} X(T^j\omega) \quad a.s.$$

Using this inclusion and taking a Castaing representation of $E^{\mathcal{I}}X$ we can assert that[1]

$$E^{\mathcal{I}}X(\omega) \subset s\text{-}li\frac{1}{n}\sum_{j=0}^{n-1} X(T^j\omega) \quad a.s.$$

Claim (3):

(6.1.3)

$$\lim_{n\to\infty} \delta^*(x^*, \frac{1}{n}\sum_{j=0}^{n-1} X(T^j\omega)) = \delta^*(x^*, E^{\mathcal{I}}X(\omega)) \quad a.s. \quad \forall x^* \in \overline{B}_{E^*}.$$

(6.1.4) $$\lim_{n\to\infty} d(x, \frac{1}{n}\sum_{j=0}^{n-1} X(T^j\omega)) = d(x, E^{\mathcal{I}}X(\omega)) \quad a.s \quad \forall x \in E.$$

There is a negligible set N_0 such that for each $\omega \in \Omega \setminus N_0$

$$\sup_{n\in\mathbf{N}} \frac{1}{n}\sum_{j=0}^{n-1} |X|(T^j\omega) < \infty$$

and there is a negligible set N_k such that for each $\omega \in \Omega \setminus N_k$

$$\lim_{n\to\infty} \delta^*(e_k, \frac{1}{n}\sum_{j=0}^{n-1} X(T^j\omega)) = \delta^*(e_k, E^{\mathcal{I}}X(\omega))$$

and also there is a negligible set M such that for each $\omega \in \Omega \setminus M$

$$E^{\mathcal{I}}X(\omega) \subset s\text{-}li\frac{1}{n}\sum_{j=0}^{n-1} X(T^j\omega).$$

Then $N = \cup_{k\geq 0} N_k \cup M$ is negligible. Let $\omega \in \Omega\setminus N$, $e^* \in D_1^*$ and $x^* \in \overline{B}_{E^*}$. We have the estimate

$$\begin{aligned}|\delta^*(x^*, \frac{1}{n}\sum_{j=0}^{n-1} X(T^j\omega)) - \delta^*(x^*, E^{\mathcal{I}}X(\omega))| \leq ||x^* - e^*|| \sup_{n\in\mathbf{N}} \frac{1}{n}\sum_{j=0}^{n-1} |X|(T^j\omega)\\ +|\delta^*(e^*, \frac{1}{n}\sum_{j=0}^{n-1} X(T^j\omega)) - \delta^*(e^*, E^{\mathcal{I}}X(\omega))|\\ +||x^* - e^*|| E^{\mathcal{I}}|X|(\omega).\end{aligned}$$

[1] This argument is classical, see the proof of Theorem 4.6 in [8].

Then from the preceding estimate, it is immediate to see that for $\omega \in \Omega \setminus N$ and $x^* \in \overline{B}_{E^*}$ we have

$$\lim_{n\to\infty} \delta^*(x^*, \frac{1}{n}\sum_{j=0}^{n-1} X(T^j\omega)) = \delta^*(x^*, E^{\mathcal{I}}X(\omega))$$

which proves (6.1.3). Let $\omega \in \Omega \setminus N$ we have

$$E^{\mathcal{I}}X(\omega) \subset s\text{-}li \frac{1}{n}\sum_{j=0}^{n-1} X(T^j\omega)$$

so that

$$\limsup_{n\to\infty} d(x, \frac{1}{n}\sum_{j=0}^{n-1} X(T^j\omega)) \le d(x, E^{\mathcal{I}}X(\omega)) \quad \forall x \in E.$$

Further for $(\omega, x) \in \Omega \setminus N \times E$ we have

$$(6.1.5) \qquad \liminf_{n\to\infty} d(x, \tfrac{1}{n}\sum_{j=0}^{n-1} X(T^j\omega))$$

$$= \liminf_{n\to\infty} \sup_{x^* \in \overline{B}^*_{E^*}} [\langle x^*, x\rangle - \delta^*(x^*, \frac{1}{n}\sum_{j=0}^{n-1} X(T^j\omega))]$$

$$\ge \sup_{k\in\mathbf{N}} \lim_{n\to\infty} [\langle e_k^*, x\rangle - \delta^*(e_k^*, \frac{1}{n}\sum_{j=0}^{n-1} X(T^j\omega))]$$

$$= \sup_{k\in\mathbf{N}} [\langle e_k^*, x\rangle - \delta^*(e_k^*, E^{\mathcal{I}}X(\omega))]$$

$$= d(x, E^{\mathcal{I}}X(\omega)).$$

Note that (6.1.5) follows also from Claim (1) by taking $B = \{x\}$. Hence for $(\omega, x) \in \Omega \setminus N \times E$ we have

$$\lim_{n\to\infty} d(x, \frac{1}{n}\sum_{j=0}^{n-1} X(T^j\omega)) = d(x, E^{\mathcal{I}}X(\omega))$$

which proves (6.1.4). Formally using (6.1.4) we get

$$(6.1.6) \qquad \limsup_{n\to\infty} D(B, \frac{1}{n}\sum_{j=0}^{n-1} X(T^j\omega)) \le D(B, E^{\mathcal{I}}X(\omega)) \quad a.s.$$

for all closed bounded convex set $B \subset E$ because we have

$$\begin{aligned}\limsup_{n\to\infty} D(B, \frac{1}{n}\sum_{j=0}^{n-1} X(T^j\omega)) &= \limsup_{n\to\infty}\inf_{x\in B} d(x, \frac{1}{n}\sum_{j=0}^{n-1} X(T^j\omega)) \\ &\le \inf_{x\in B}\limsup_{n\to\infty} d(x, \frac{1}{n}\sum_{j=0}^{n-1} X(T^j\omega)) \\ &= \inf_{x\in B} d(x, E^{\mathcal{I}}X(\omega)) = D(B, E^{\mathcal{I}}X(\omega)).\end{aligned}$$

So combining (6.1.6) with Claim (1) we get

$$\lim_{n\to\infty} D(B, \frac{1}{n}\sum_{j=0}^{n-1} X(T^j\omega)) = D(B, E^{\mathcal{I}}X(\omega)) \quad a.s.$$

for all closed bounded convex set B of E. ■

Remark. Since $X \in \mathcal{L}^1_{cwk(E)}(\Omega, \mathcal{F}, P)$, and the strong dual E^*_b is separable, by [33], $E^{\mathcal{B}}X \in \mathcal{L}^1_{cwk(E)}(\mathcal{I})$ with $\mathcal{S}^1_{E^{\mathcal{I}}X}(\mathcal{I}) = \{E^{\mathcal{B}}f : f \in \mathcal{S}^1_X(\mathcal{F})\}$. Using this special property of the $E^{\mathcal{I}}X$, it is easy to prove in inclusion

$$w\text{-}ls\frac{1}{n}\sum_{j=0}^{n-1} X(T^j\omega) \subset E^{\mathcal{I}}X(\omega) \quad a.s.$$

Let $x \in w\text{-}ls S_n(\omega)$ with $\omega \in \Omega \setminus N$, there is $x_p \to x$ weakly for some $x_p \in S_{n_p}(\omega)$ $(p \ge 1)$. As

$$\langle e^*_k, x\rangle = \lim_{p\to\infty}\langle e^*_k, x_p\rangle \le \lim_{p\to\infty}\delta^*(e^*_k, S_{n_p}(\omega)) = \delta^*(e^*_k, E^{\mathcal{I}}X(\omega)) \quad \forall k \in \mathbf{N}$$

we get $x \in E^{\mathcal{I}}X(\omega)$ because $E^{\mathcal{I}}X(\omega)$ is convex weakly compact (see [16], Proposition III.35). From this inclusion and Claim (2) we have

$$\text{Mosco - }\lim_{n\to\infty}\frac{1}{n}\sum_{j=0}^{n-1} X(T^j\omega) = E^{\mathcal{I}}X(\omega) \quad a.s.$$

It is worth to address the question of validity of Theorem 6.1 when the strong dual is no longer separable. For this purpose we will provide a variant of Theorem 6.1 by using some compactness assumptions. Nevertheless this need a careful look. A convex weakly compact valued measurable mapping $X : \Omega \to cwk(E)$ is *weak compactly integrably bounded* if there exist $\alpha \in L^1_{\mathbf{R}^+}(\Omega, \mathcal{F}, P)$ and a convex weakly compact set K in E such that $X(\omega) \subset \alpha(\omega)K$ for all $\omega \in \Omega$. It is obvious that "weak compactly integrably bounded" implies "integrably bounded".

Theorem 6.2. *Assume that E is separable, T is a measurable transformation of Ω preserving P, $\mathcal{I}$ is the σ-algebra of invariant sets and $X : \Omega \to cwk(E)$ is a $\mathcal{F}$-measurable and weak compactly integrably bounded mapping, then the following hold:*

$$\lim_{n\to\infty} \delta^*(x^*, \frac{1}{n}\sum_{j=0}^{n-1} X(T^j\omega)) = \delta^*(x^*, E^{\mathcal{I}}X(\omega))$$

for all $x^* \in \overline{B}_{E^*}$ and almost surely $\omega \in \Omega$, and

$$\lim_{n\to\infty} d(x, \frac{1}{n}\sum_{j=0}^{n-1} X(T^j\omega)) = d(x, E^{\mathcal{I}}X(\omega))$$

for all $x \in E$ and almost surely $\omega \in \Omega$, and consequently

$$\lim_{n\to\infty} D(B, \frac{1}{n}\sum_{j=0}^{n-1} X(T^j\omega)) = D(B, E^{\mathcal{I}}X(\omega))$$

for all convex weakly compact subsets B of E and almost surely $\omega \in \Omega$. Here $E^{\mathcal{I}}X$ is the conditional expectation of X in the sense of Hiai-Umegaki.

Proof. By assumption there exist $\alpha \in L^1_{\mathbf{R}^+}(\Omega, \mathcal{F}, P)$ and a convex weakly compact set K in E such that $X(\omega) \subset \alpha(\omega)K$ for all $\omega \in \Omega$. W.l.o.g we may assume that K is equilibrated. It is not difficult to see that $E^{\mathcal{I}}X(\omega) \subset E^{\mathcal{I}}\alpha(\omega)K$ for all $\omega \in \Omega$. Hence $E^{\mathcal{I}}X(\omega)$ is convex weakly compact. Let $D_1^* = (e_k^*)_{k\in\mathbf{N}}$ a dense sequence in the closed unit ball $\overline{B}_{E^*}$ with respect to the Mackey topology. Next, applying the classical Birkhoff ergodic theorem to α and each $\delta^*(e_k^*, X)$ yields

$$\lim_{n\to\infty} \frac{1}{n}\sum_{j=0}^{n-1} \alpha(T^j\omega) = E^{\mathcal{I}}\alpha(\omega) \quad a.s. \tag{6.2.1}$$

$$\begin{aligned}\lim_{n\to\infty} \delta^*(e_k^*, \frac{1}{n}\sum_{j=0}^{n-1} X(T^j\omega)) &= \lim_{n\to\infty} \frac{1}{n}\sum_{j=0}^{n-1} \delta^*(e_k^*, X(T^j\omega))\\ &= \delta^*(e_k^*, E^{\mathcal{I}}X(\omega)) \quad a.s. \end{aligned} \tag{6.2.2}$$

Consequently by (6.2.1) $\frac{1}{n}\sum_{j=0}^{n-1}\alpha(T^j\omega)$ is pointwise bounded a.s., say

$$\beta(\omega) := \sup_{n\in\mathbf{N}} \frac{1}{n}\sum_{j=0}^{n-1}\alpha(T^j\omega) < \infty \quad a.s.$$

It follows that

$$S_n(\omega) := \frac{1}{n}\sum_{j=0}^{n-1} X(T^j\omega) \subset [\frac{1}{n}\sum_{j=0}^{n-1}\alpha(T^j\omega)]K \subset \beta(\omega)K \quad a.s.$$

Claim (1):

$$\liminf_{n\to\infty} D(B, \frac{1}{n}\sum_{j=0}^{n-1} X(T^j\omega)) \geq D(B, E^{\mathcal{I}}X(\omega))$$

for all convex weakly compact subset $B \subset E$ and almost surely $\omega \in \Omega$. Indeed, for each $k \in \mathbf{N}$, there is a negligible set N_k such that (6.2.2) holds. Then $N := \cup_k N_k$ is negligible and for $\omega \in \Omega \setminus N$ we have

$$\liminf_{n\to\infty} D(B, \frac{1}{n}\sum_{j=0}^{n-1} X(T^j\omega))$$

$$= \liminf_{n\to\infty} \sup_{x\in\overline{B}_{E^*}} [-\delta^*(x^*, \frac{1}{n}\sum_{j=0}^{n-1} X(T^j\omega)) - \delta^*(-x^*, B)|$$

$$\geq \sup_{x^*\in D_1^*} \liminf_{n\to\infty} [-\delta^*(x^*, \frac{1}{n}\sum_{j=0}^{n-1} X(T^j\omega)) - \delta^*(-x^*, B)]$$

$$= \sup_{x^*\in D_1^*} [-\limsup_{n\to\infty} \delta^*(x^*, \frac{1}{n}\sum_{j=0}^{n-1} X(T^j\omega)) - \delta^*(-x^*, B)]$$

$$= \sup_{x^*\in D_1^*} [-\delta^*(x^*, E^{\mathcal{I}}X(\omega)) - \delta^*(-x^*, B)] = D(B, E^{\mathcal{I}}X(\omega)).$$

Claim (2):

$$E^{\mathcal{I}}X(\omega) \subset s\text{-}li\frac{1}{n}\sum_{j=0}^{n-1} X(T^j\omega) \quad a.s.$$

Recall that

$$S^1_{E^{\mathcal{I}}X} = cl\{E^{\mathcal{I}}f : f \in S^1_X\}$$

here the *closure* is taken in the sense of the norm in L^1_E (see also Theorem 3 in [33]). Firstly we check that

$$S^1_{E^{\mathcal{I}}X} \subset S^1_{s\text{-}li\frac{1}{n}\sum_{j=0}^{n-1} X(T^j)}$$

Let $f \in \mathcal{S}^1_X$. Applying Birkhoff ergodic theorem for Banach valued functions (see [26], Theorem 2.1, p.167) yields

$$\lim_{n\to\infty} \frac{1}{n}\sum_{j=0}^{n-1} f(T^j\omega) = E^{\mathcal{I}} f(\omega) \quad a.s.$$

Whence for a.s. $\omega \in \Omega$, $E^{\mathcal{I}} f(\omega)$ is the strong limit of $\frac{1}{n}\sum_{j=0}^{n-1} f(T^j\omega) \in S_n(\omega) := \frac{1}{n}\sum_{j=0}^{n-1} X(T^j\omega)$. In other words

$$E^{\mathcal{I}} f(\omega) \in s\text{-}li\frac{1}{n}\sum_{j=0}^{n-1} X(T^j\omega) \quad a.s.$$

Now if $g \in S^1_{E^{\mathcal{I}}X}$. There is a sequence (f_n) in $\mathcal{S}^1_X$ such that $E^{\mathcal{I}} f_n \to g$ a.s. so that

$$g(\omega) \in s\text{-}li\frac{1}{n}\sum_{j=0}^{n-1} X(T^j\omega) \quad a.s.$$

Using this inclusion and taking a Castaing representation of $E^{\mathcal{I}}X$ we can assert that

$$E^{\mathcal{I}}X(\omega) \subset s\text{-}li\frac{1}{n}\sum_{j=0}^{n-1} X(T^j\omega) \quad a.s.$$

Now we are going to prove

Claim (3):

(6.2.3)

$$\lim_{n\to\infty} \delta^*(x^*, \frac{1}{n}\sum_{j=0}^{n-1} X(T^j\omega)) = \delta^*(x^*, E^{\mathcal{I}}X(\omega)) \quad a.s. \quad \forall x^* \in \overline{B}_{E^*}.$$

(6.2.4) $$\lim_{n\to\infty} d(x, \frac{1}{n}\sum_{j=0}^{n-1} X(T^j\omega)) = d(x, E^{\mathcal{I}}X(\omega)) \quad a.s \quad \forall x \in E.$$

There is a negligible set N_0 such that for each $\omega \in \Omega \setminus N_0$

$$\beta(\omega) := \sup_{n\in\mathbf{N}} \frac{1}{n}\sum_{j=0}^{n-1} \alpha(T^j\omega) < \infty$$

and there is a negligible set N_k such that for each $\omega \in \Omega \setminus N_k$

$$\lim_{n\to\infty} \delta^*(e_k, \frac{1}{n}\sum_{j=0}^{n-1} X(T^j\omega)) = \delta^*(e_k, E^{\mathcal{I}}X(\omega))$$

and also there is a negligible set M such that for each $\omega \in \Omega \setminus M$

$$E^{\mathcal{I}}X(\omega) \subset s\text{-}li \frac{1}{n}\sum_{j=0}^{n-1} X(T^j\omega).$$

Then $N = \cup_{k\geq 0} N_k \cup M$ is negligible. Let $\omega \in \Omega \setminus N$, $x^* \in \overline{B}_{E^*}$ and $\varepsilon > 0$ Pick $e_j^* \in D_1^*$ such that

$$\max\{\delta^*(x^* - e_j^*, E^{\mathcal{I}}\alpha(\omega)K), \delta^*(e_j^* - x^*, E^{\mathcal{I}}\alpha(\omega)K)\} < \varepsilon$$

and

$$\max\{\delta^*(x^* - e_j^*, \beta(\omega)K), \delta^*(e_j^* - x^*, \beta(\omega)K)\} < \varepsilon.$$

Let us write

$$\begin{aligned} &|\delta^*(x^*, S_n(\omega)) - \delta^*(x^*, E^{\mathcal{I}}X(\omega))| \\ &= |\delta^*(x^*, S_n(\omega)) - \delta^*(e_j^*, S_n(\omega))| \\ &+|\delta^*(e_j^*, S_n(\omega)) - \delta^*(e_j^*, E^{\mathcal{I}}X(\omega))| \\ &+|\delta^*(e_j^*, E^{\mathcal{I}}X(\omega)) - \delta^*(x^*, E^{\mathcal{I}}X(\omega))|. \end{aligned}$$

As $S_n(\omega) \subset \beta(\omega)K$ and $E^{\mathcal{I}}X(\omega) \subset E^{\mathcal{I}}\alpha(\omega)K$ for all $n \in \mathbf{N}$ and for all $\omega \in \Omega \setminus N$, we have the estimates

$$\begin{aligned} &|\delta^*(x^*, S_n(\omega)) - \delta^*(e_j^*, S_n(\omega))| \\ &\leq \max\{\delta^*(x^* - e_j^*, \beta(\omega)K), \delta^*(e_j^* - x^*, \beta(\omega)K)\} < \varepsilon \end{aligned}$$

and

$$\begin{aligned} &|\delta^*(e_j^*, E^{\mathcal{I}}X(\omega)) - \delta^*(x^*, E^{\mathcal{I}}X(\omega))| \\ &\leq \max\{\delta^*(x^* - e_j^*, E^{\mathcal{I}}\alpha(\omega)K), \delta^*(e_j^* - x^*, E^{\mathcal{I}}\alpha(\omega)K)\} < \varepsilon. \end{aligned}$$

Finally we get

$$[\delta^*(x^*, S_n(\omega)) - \delta^*(x^*, E^{\mathcal{I}}X(\omega))] < [\delta^*(e_j^*, S_n(\omega)) - \delta^*(e_j^*, E^{\mathcal{I}}X(\omega))] + 2\varepsilon.$$

As $[\delta^*(e_j^*, S_n(\omega)) - \delta^*(e_j^*, E^{\mathcal{I}}X(\omega))] \to 0$, from the preceding estimate, it is immediate to see that for $\omega \in \Omega \setminus N$ and $x^* \in \overline{B}_{E^*}$ we have

$$\lim_{n\to\infty} \delta^*(x^*, \frac{1}{n}\sum_{j=0}^{n-1} X(T^j\omega)) = \delta^*(x^*, E^{\mathcal{I}}X(\omega))$$

which proves (6.2.3). Let $\omega \in \Omega \setminus N$ we have

$$E^{\mathcal{I}}X(\omega)) \subset s\text{-}li \frac{1}{n}\sum_{j=0}^{n-1} X(T^j\omega)$$

so that

$$\limsup_{n\to\infty} d(x, \frac{1}{n}\sum_{j=0}^{n-1} X(T^j\omega)) \le d(x, E^{\mathcal{I}}X(\omega)) \quad \forall x \in E.$$

Further for $(\omega, x) \in \Omega \setminus N \times E$ we have

(6.2.5) $$\liminf_{n\to\infty} d(x, \tfrac{1}{n}\sum_{j=0}^{n-1} X(T^j\omega))$$

$$= \liminf_{n\to\infty} \sup_{x^*\in \overline{B}^*_{E^*}} [\langle x^*, x\rangle - \delta^*(x^*, \frac{1}{n}\sum_{j=0}^{n-1} X(T^j\omega))]$$

$$\ge \sup_{k\in\mathbf{N}} \lim_{n\to\infty} [\langle e_k^*, x\rangle - \delta^*(e_k^*, \frac{1}{n}\sum_{j=0}^{n-1} X(T^j\omega))]$$

$$= \sup_{k\in\mathbf{N}} [\langle e_k^*, x\rangle - \delta^*(e_k^*, E^{\mathcal{I}}X(\omega))]$$

$$= d(x, E^{\mathcal{I}}X(\omega)).$$

Note that (6.2.5) follows also from Claim (1) by taking $B = \{x\}$. Hence for $(\omega, x) \in \Omega \setminus N \times E$ we have

$$\lim_{n\to\infty} d(x, \frac{1}{n}\sum_{j=0}^{n-1} X(T^j\omega)) = d(x, E^{\mathcal{I}}X(\omega))$$

which proves (6.2.4). Formally using (6.2.4) we get

$$\limsup_{n\to\infty} D(B, \frac{1}{n}\sum_{j=0}^{n-1} X(T^j\omega)) \leq D(B, E^{\mathcal{I}}X(\omega)) \quad a.s. \tag{6.2.6}$$

for all closed bounded convex set $B \subset E$ because we have

$$\begin{aligned}\limsup_{n\to\infty} D(B, \frac{1}{n}\sum_{j=0}^{n-1} X(T^j\omega)) &= \limsup_{n\to\infty}\inf_{x\in B} d(x, \frac{1}{n}\sum_{j=0}^{n-1} X(T^j\omega)) \\ &\leq \inf_{x\in B}\limsup_{n\to\infty} d(x, \frac{1}{n}\sum_{j=0}^{n-1} X(T^j\omega)) \\ &= \inf_{x\in B} d(x, E^{\mathcal{I}}X(\omega)) = D(B, E^{\mathcal{I}}X(\omega)).\end{aligned}$$

So combining (6.2.6) with Claim (1) we get

$$\lim_{n\to\infty} D(B, \frac{1}{n}\sum_{j=0}^{n-1} X(T^j\omega)) = D(B, E^{\mathcal{I}}X(\omega)) \quad a.s.$$

for all convex weakly compact set B of E. ∎

For more information on the conditional expectation of multifunctions, we refer to [2, 9, 20, 33]. In particular, recent existence results for conditional expectation in Gelfand and Pettis integration as well as the multivalued Dunford-Pettis representation theorem are available [2, 9]. These results involve several new convergence problems, for instance, the Mosco convergence of sub-super martingales, pramarts in Bochner, Pettis or Gelfand integration, see [2, 9, 10, 12].

The following variant deals with Gelfand integration, here E^* is no longer strongly separable. A mapping $\Gamma : \Omega \to cwk(E^*_s)$ is Gelfand-integrable, if the mapping $\delta^*(e, \Gamma(.))$ is integrable, for all $e \in E$. The Aumann-Gelfand integral of Γ, denoted by

$$AG\text{-}\int_\Omega G dP = \{G\text{-}\int_\Omega f dP : f \in \mathcal{S}_\Gamma\}$$

where $\mathcal{S}_\Gamma$ is the set of all Gelfand-integrable selections of Γ and $G\text{-}\int_\Omega f dP$ is the Gelfand integral of $f \in \mathcal{S}_\Gamma$. We need to recall the following result on the existence of conditional expectation of convex weak star compact valued Gelfand-integrable mapping ([9], Theorem 6.1).

Theorem 6.3. *Let $\mathcal{B}$ be a sub-σ-algebra of $\mathcal{F}$ and let X be a $cwk(E^*_s)$-valued Gelfand-integrable mapping such that $E^{\mathcal{B}}|X| \in [0, +\infty[$. Then there* *exists a unique $\mathcal{B}$-measurable, $cwk(E^*_s)$-valued Gelfand-integrable mapping, denoted by Ge-$E^{\mathcal{B}}X$ which enjoys the following property: For every $h \in L^\infty(\mathcal{B})$, one has*

$$(*) \qquad AG\text{-}\int_\Omega h Ge\text{-}E^{\mathcal{B}} X dP = AG\text{-}\int_\Omega h X dP.$$

Ge-$E^{\mathcal{B}}X$ is called the Gelfand conditional expectation of X.

We need the following definition.

Definition 6.1. *The Banach space E is weakly compact generated (WCG) if there exists a weakly compact subset of E whose linear span is dense in E.*

Theorem 6.4. *Let E be a separable WCG Banach space, T a measurable transformation of Ω preserving P, $\mathcal{I}$ the σ-algebra of invariant sets and $X : \Omega \to cwk(E^*)$ a $\mathcal{F}$-measurable and Gelfand-integrable mapping such that $E^{\mathcal{I}}|X| \in [0, +\infty[$. Then the following hold:*

$$w^*\text{-Kuratowski - }\lim_{n\to\infty} \frac{1}{n}\sum_{j=0}^{n-1} X(T^j\omega) = E^{\mathcal{I}}X(\omega) \quad a.s.$$

where $E^{\mathcal{I}}X := Ge\text{-}E^{\mathcal{I}}X$ (for short) denotes the Gelfand conditional expectation of X.

Proof. Let $D_1 = (e_k)_{k\in\mathbf{N}}$ be a dense sequence in $\overline{B}_E$. Note that the mapping $|X|$

$$|X|(\omega) = \sup_{x\in\overline{B}_E} |\delta^*(x, X(\omega))| \quad \forall\omega \in \Omega$$

is $\mathcal{F}$-measurabie (see e.g. [16] or [2]). Next, applying the classical real valued Birkhoff ergodic theorem to $|X|$ and each $\delta^*(e_k, X)$ yields

$$(6.4.1) \qquad \lim_{n\to\infty} \frac{1}{n}\sum_{j=0}^{n-1} |X|(T^j\omega) = E^{\mathcal{I}}|X|(\omega) \quad a.s.$$

$$\lim_{n\to\infty} \delta^*(e_k, \frac{1}{n}\sum_{j=0}^{n-1} X(T^j\omega)) = \lim_{n\to\infty} \frac{1}{n}\sum_{j=0}^{n-1} \delta^*(e_k, X(T^j\omega))$$

$$(6.4.2) \qquad = \delta^*(e_k, E^{\mathcal{I}}X(\omega)) \quad a.s.$$

Consequently there is a negligible set N_0 such that for each $\omega \in \Omega \setminus N_0$

$$(6.4.3) \qquad \sup_{n\in\mathbf{N}} \frac{1}{n}\sum_{j=0}^{n-1} |X|(T^j\omega) < \infty$$

and there is a negligible set N_k such that for each $\omega \in \Omega \setminus N_k$

(6.4.4) $$\lim_{n\to\infty} \delta^*(e_k, \frac{1}{n}\sum_{j=0}^{n-1} X(T^j\omega)) = \delta^*(e_k, E^{\mathcal{I}}X(\omega)).$$

Claim

$$w^*\text{-Kuratowski - }\lim_{n\to\infty} \frac{1}{n}\sum_{j=0}^{n-1} X(T^j\omega) = E^{\mathcal{I}}X(\omega) \quad a.s.$$

Set $N = \cup_{k\geq 0} N_k$. Then N is negligible. Let $\omega \in \Omega \setminus N$, $e \in D_1$ and $x \in \overline{B}_E$. We have the estimate

$$|\delta^*(x, \frac{1}{n}\sum_{j=0}^{n-1} X(T^j\omega)) - \delta^*(x, E^{\mathcal{I}}X(\omega))| \leq ||x-e|| \sup_{n\in\mathbf{N}} \frac{1}{n}\sum_{j=0}^{n-1} |X|(T^j\omega)$$

$$+|\delta^*(e, \frac{1}{n}\sum_{j=0}^{n-1} X(T^j\omega)) - \delta^*(e, E^{\mathcal{I}}X(\omega))|$$

$$+||x-e|| E^{\mathcal{I}}|X|(\omega).$$

Then from (6.4.3) and (6.4.4) and the preceding estimate, it is immediate to see that for $\omega \in \Omega \setminus N$ and $x \in \overline{B}_E$ we have

(6.4.5) $$\lim_{n\to\infty} \delta^*(x, \frac{1}{n}\sum_{j=0}^{n-1} X(T^j\omega)) = \delta^*(x, E^{\mathcal{I}}X(\omega)).$$

Now since E is WCG, from (6.4.3)-(6.4.4)-(6.4.5) and ([19], Theorem 4.1) we deduce that

$$w^*\text{-Kuratowski - }\lim_{n\to\infty} \frac{1}{n}\sum_{j=0}^{n-1} X(T^j\omega) = E^{\mathcal{I}}X(\omega) \quad a.s.$$

■

Now we treat the multivalued ergodic theorem in Pettis integration. A mapping $\Gamma : \Omega \to cwk(E)$ is Pettis-integrable, if the mapping $\delta^*(e^*, \Gamma(.))$ is integrable for all $e^* \in E^*$ and if any scalarly integrable selection of Γ is Pettis-integrable. The Aumann-Pettis integral of Γ is defined by

$$AP\text{-}\int_\Omega \Gamma dP = \{P\text{-}\int_\Omega f dP : f \in \mathcal{S}_\Gamma^{Pe}\}$$

where ${\mathcal{S}^{Pe}}_\Gamma$ is the set of all Pettis-integrable selections of Γ and $P\text{-}\int_\Omega f dP$ is the Pettis integral of $f \in \mathcal{S}_\Gamma^{Pe}$. We need to recall the following result on the existence of conditional expectation of convex weak star compact valued Pettis-integrable mapping ([2], Theorem 4.4).

Theorem 6.5. *Assume that E_b^* is separable. Let $\mathcal{B}$ be a sub-σ-algebra of $\mathcal{F}$ and let X be a $cwk(E)$-valued Pettis-integrable mapping such that $E^{\mathcal{B}}|X| \in [0, +\infty[$. Then there exists a unique $\mathcal{B}$-measurable, $cwk(E)$-valued Pettis-integrable multifunction, denoted by $Pe\text{-}E^{\mathcal{B}}X$, which enjoys the following property: For every $h \in L^\infty(\mathcal{B})$, one has*

$$Pe\text{-}\int_\Omega h Pe\text{-}E^{\mathcal{B}} X dP = Pe\text{-}\int_\Omega hX dP$$

where $Pe\text{-}\int_\Omega h Pe\text{-}E^{\mathcal{B}} X dP$ and $Pe\text{-}\int_\Omega hX dP$ denote the $cwk(E)$-valued Aumann-Pettis integral of $h Pe\text{-}E^{\mathcal{B}} X$ and hX respectively.

Theorem 6.6. *Assume that E and its strong dual E_b^* are separable, T is a measurable transformation of Ω preserving P, $\mathcal{I}$ is the σ-algebra of invariant sets and $X : \Omega \to cwk(E)$ is a $\mathcal{F}$-measurable and Pettis-integrable mapping such that $E^{\mathcal{I}}|X| \in [0, +\infty[$.*
Then the following hold:

$$\lim_{n\to\infty} \delta^*(x^*, \frac{1}{n}\sum_{j=0}^{n-1} X(T^j\omega)) = \delta^*(x^*, E^{\mathcal{I}}X(\omega))$$

for all $x^* \in \overline{B}_{E^*}$ and almost surely $\omega \in \Omega$, where $E^{\mathcal{I}}X := Pe\text{-}E^{\mathcal{I}}X$ (for short) denotes the Pettis conditional expectation of X.

Proof. Let $D_1^* = (e_k^*)_{k\in\mathbf{N}}$ a dense sequence in $\overline{B}_{E^*}$ with respect to the norm dual topology. Note that the mapping $|X|$

$$|X|(\omega) = \sup_{x^*\in\overline{B}_{E^*}} |\delta^*(x^*, X(\omega))| \quad \forall\omega \in \Omega$$

is $\mathcal{F}$-measurabile (see e.g. [16]). Next, applying the classical Birkhoff ergodic theorem to $|X|$ and each $\delta^*(e_k^*, X)$ (see e.g. [34]) yields almost surely

(6.6.1) $$\lim_{n\to\infty} \frac{1}{n}\sum_{j=0}^{n-1} |X|(T^j\omega) = E^{\mathcal{I}}|X|(\omega) \quad a.s.$$

$$\lim_{n\to\infty} \delta^*(e_k^*, \frac{1}{n}\sum_{j=0}^{n-1} X(T^j\omega)) = \lim_{n\to\infty} \frac{1}{n}\sum_{j=0}^{n-1} \delta^*(e_k^*, X(T^j\omega))$$

(6.6.2) $$= \delta^*(e_k^*, E^{\mathcal{I}}X(\omega)).$$

Consequently there is a negligible set N_0 such that for each $\omega \in \Omega \setminus N_0$

$$R(\omega) := \sup_{n \in \mathbf{N}} \frac{1}{n} \sum_{j=0}^{n-1} |X|(T^j \omega) < \infty \tag{6.6.3}$$

and there is a negligible set N_k such that for each $\omega \in \Omega \setminus N_k$

$$\lim_{n\to\infty} \delta^*(e_k^*, \frac{1}{n} \sum_{j=0}^{n-1} X(T^j \omega)) = \delta^*(e_k^*, E^{\mathcal{I}} X(\omega)) \tag{6.6.4}$$

Set $N = \cup_{k \geq 0} N_k$. Then N is negligible. Let $\omega \in \Omega \setminus N$, $x^* \in \overline{B}_{E^*}$ and $e^* \in D_1^*$. By (6.6.3) we have

$$|S_n(\omega)| \leq R(\omega) < \infty$$

for each $\omega \in \Omega \setminus N$. We have the estimate

$$\begin{aligned} |\delta^*(x,^*, \frac{1}{n} \sum_{j=0}^{n-1} X(T^j \omega)) - \delta^*(x^*, E^{\mathcal{I}} X(\omega))| &\leq ||x^* - e^*|| R(\omega) \\ &+ |\delta^*(e^*, \frac{1}{n} \sum_{j=0}^{n-1} X(T^j \omega)) - \delta^*(e^*, E^{\mathcal{I}} X(\omega))| \\ &+ ||x^* - e^*|| E^{\mathcal{I}} |X|(\omega). \end{aligned}$$

Then from (6.6.3) and (6.6.4) and the preceding estimate, it is immediate to see that for $\omega \in \Omega \setminus N$ and $x^* \in \overline{B}_{E^*}$ we have

$$\lim_{n\to\infty} \delta^*(x^*, \frac{1}{n} \sum_{j=0}^{n-1} X(T^j \omega)) = \delta^*(x^*, E^{\mathcal{I}} X(\omega)). \qquad \blacksquare$$

7. An unusual convergence for superadditive integrable process in Banach lattice

We must recall some facts from ergodic theory and provide the definition of superadditive process in $L^1_{\mathbf{R}}(\Omega, \mathcal{F}, P)$. A sequence $(S_m)_{m \in \mathbf{N}}$ in $L^1_{\mathbf{R}}(\Omega, \mathcal{F}, P)$ is a superadditive process provided that, for all $m, n \in \mathbf{N}$, we have

$$S_{m+n}(\omega) \geq S_m(\omega) + S_n(T^m \omega)$$

for all $\omega \in \Omega$.

Theorem (J.F.C. Kingmann) [18]. *For any superadditive sequence* $(S_n)_{n\in\mathbf{N}}$ *of integrable random variable,* $(\frac{S_n}{n})_{n\in\mathbf{N}}$ *converges a.s. as* $n \to \infty$ *to* $\gamma := \sup_{n\in\mathbf{N}} \frac{E^{\mathcal{I}} S_n}{n}$. *Here* γ *is integrable if and only if* $\sup_{n\in\mathbf{N}} \frac{E(S_n)}{n} < \infty$, *and if* γ *is integrable, then* $(\frac{S_n}{n})_{n\in\mathbf{N}}$ *converges to* γ *also in* L^1.

In the following let E denotes a separable Banach lattice and E^*_+ denotes the positive cone in the dual E^*. For background on properties of Banach lattices, we may consult Schafer [31], Peressini [30]. The following definition is formally similar to the case of real valued integrable superadditive process.

Definition 7.1. *Let* E *be a separable Banach lattice, let* T *a measurable transformation of* Ω *preserving* P. *A sequence* $(S_m)_{m\in\mathbf{N}}$ *in* $L^1_E(\Omega, \mathcal{F}, P)$ *is a superadditive process provided that for all* $m, n \in \mathbf{N}$, *and for all* $\omega \in \Omega$, *we have*

$$S_{m+n}(\omega) \geq S_m(\omega) + S_n(T^m\omega)$$

for all $\omega \in \Omega$.

The above superadditive condition means that for all $f^* \in E^*_+$, for all $m, n \in \mathbf{N}$, for all $\omega \in \Omega$, we have

$$\langle f^*, S_{m+n}(\omega)\rangle \geq \langle f^*, S_m(\omega) + S_n(T^m\omega)\rangle.$$

In otherwords, for each $f^* \in E^*_+$, $(\langle f^*, S_m\rangle)_{m\in\mathbf{N}}$ is a superadditive process in $L^1_{\mathbf{R}}(\Omega, \mathcal{F}, P)$.

The following is an unusual convergence of superadditive vector valued process with localization of the limit. Surprisingly this result can be considered as a by product of some results in Mathematical Economics dealing with the existence of integrable selection of the sequential weak upper limit of a sequence of measurable multifunctions. See Theorem 4.9 in [12]. At this point, let us mention some important variants. Krupa [27] and Schurger [32] treated the Ergodic theorems for subadditive superstationary families of convex compact random sets using Abid result [1] on the convergence of subadditive superstationary process while Ghoussoub and Steele [6] treated the a.s. norm convergence for subadditive process in an order complete Banach lattice extending the Kingman's theorem for real valued subadditive process.

Theorem 7.1. *Assume that* E *is separable Banach lattice,* E^*_+ *is the positive cone in the dual* E^*, T *is a measurable transformation of* Ω *preserving* P, $\mathcal{I}$ *is the* σ*-algebra of invariant sets and* $(S_n)_{n\in\mathbf{N}}$ *is a superadditive process in* $L^1_E(\Omega, \mathcal{F}, P)$ *satisfying:*

(*i*) *There is a sequence* $(r_n)_{n\in\mathbf{N}}$ *in* $L^0_{\mathbf{R}}(\Omega, \mathcal{F}, P)$ *with* $r_n \in co\{|\frac{S_i}{i}| : i \geq n\}$ *such that* $\limsup r_n \in L^1_{\mathbf{R}}(\Omega, \mathcal{F}, P)$.

(ii) *For each* $f^* \in E^*$, $(\frac{\langle f^*, S_n\rangle}{n})_{n\in\mathbf{N}}$ *is uniformly integrable.*
(iii) *There is a convex ball weakly compact valued multifunction* $\Gamma : \Omega \Rightarrow E$ *such that*

$$\frac{S_n(\omega)}{n} \in \Gamma(\omega) \quad \forall n \in \mathbf{N} \quad \forall \omega \in \Omega.$$

Then there exists $Z_\infty \in L^1_E(\Omega, \mathcal{F}, P)$ *verifying:*

(a) $\int_A \langle f^*, Z_\infty\rangle dP = \lim_{n\to} \int_A \langle f^*, \frac{S_n}{n}\rangle dP = \int_A \sup_{n\in\mathbf{N}} \langle f^*, \frac{E^{\mathcal{I}} S_n}{n}\rangle dP$ *for all* $A \in \mathcal{F}$ *and for all* $f^* \in E^*_+$.
(b) $Z_\infty(\omega) \in \bigcap_{m\geq 1} cl\, co\{\frac{S_n(\omega)}{n} : n \geq m\}$ *almost surely.*

Proof. Let $f^* \in E^*_+$. It is clear that each $(\langle f^*, S_n\rangle)_{n\in\mathbf{N}}$ is a real valued integrable superadditive process. By condition (ii) and Kingmann ergodic theorem, $\langle f^*, \frac{S_n}{n}\rangle$ converges a.s. and also in L^1 to the integrable function

$$\sup_{n\in\mathbf{N}} \frac{E^{\mathcal{I}}\langle f^*, S_n\rangle}{n} = \sup_{n\in\mathbf{N}} \langle f^*, \frac{E^{\mathcal{I}} S_n}{n}\rangle$$

which entails

$$\lim_{n\to\infty} \langle f^*, \int_A \frac{S_n}{n} dP\rangle = \lim_{n\to\infty} \int_A \langle f^*, \frac{S_n}{n}\rangle dP \tag{7.1}$$

(7.1.1) $$= \int_A \sup_{n\in\mathbf{N}} \langle f^*, \frac{E^{\mathcal{I}} S_n}{n}\rangle dP \quad \forall A \in \mathcal{F}.$$

Now using $(i)(ii)(iii)$ and repeating the arguments of the proof of Theorem 4.9 in [12] we provide a sequence $(Z_n)_{n\in\mathbf{N}}$ whose members are convex combinations of $(\frac{S_n}{n})$ and $Z_\infty \in L^1_E(\Omega, \mathcal{F}, P)$ such that

(7.1.2) $$weak\text{-}\lim_{n\to\infty} Z_n(\omega) = Z_\infty(\omega) \quad a.s.$$

Using (7.1.2) we get (b)

$$Z_\infty(\omega) \in \bigcap_{m\geq 1} cl\, co\{\frac{S_n(\omega)}{n} : n \geq m\} \quad a.s.$$

By (ii) we note that the sequence $(\langle f^*, Z_n\rangle)_{n\in\mathbf{N}}$ is uniformly integrable, for each $f^* \in E^*$, which entails

(7.1.3) $$\int_A \langle f^*, Z_\infty\rangle dP = \lim_{n\to\infty} \int_A \langle f^*, Z_n\rangle dP$$

for all $A \in \mathcal{F}$. Now let $f^* \in E^*_+$. Coming back to (7.1.1)- (7.1.2)-(7.1.3) we get

$$\int_A \langle f^*, Z_\infty \rangle dP = \lim_{n\to\infty} \int_A \langle f^*, Z_n \rangle dP = \lim_{n\to\infty} \langle f^*, \int_A Z_n dP \rangle$$

$$= \lim_{n\to\infty} \langle f^*, \int_A \frac{S_n}{n} dP \rangle = \lim_{n\to\infty} \int_A \langle f^*, \frac{S_n}{n} \rangle dP = \int_A \sup_{n\in\mathbf{N}} \langle f^*, \frac{E^{\mathcal{I}} S_n}{n} \rangle dP.$$

thus proving (a). ■

Here is a corollary of Theorem 7.1.

Corollary 7.1. *Assume that E is reflexive separable Banach lattice, T is a measurable transformation of Ω preserving P, $\mathcal{I}$ is the σ-algebra of invariant sets and $(S_n)_{n\in\mathbf{N}}$ is a superadditive process in $L^1_E(\Omega, \mathcal{F}, P)$ satisfying:*

(i) $\sup_{n\in\mathbf{N}} \frac{E(|S_n|)}{n} < \infty$.
(ii) For each $f^* \in E^*$, $(\frac{\langle f^*, S_n \rangle}{n})_{n\in\mathbf{N}}$ *is uniformly integrable.*

Then there exists $Z_\infty \in L^1_E(\Omega, \mathcal{F}, P)$ verifying:

(a) $\int_A \langle f^*, Z_\infty \rangle dP = \lim_{n\to} \int_A \langle f^*, \frac{S_n}{n} \rangle dP = \int_A \sup_{n\in\mathbf{N}} \langle f^*, \frac{E^{\mathcal{I}} S_n}{n} \rangle dP$ *for all $A \in \mathcal{F}$ and for all $f^* \in E^*_+$.*
(b) $Z_\infty(\omega) \in \bigcap_{m\geq 1} cl\, co\{\frac{S_n(\omega)}{n} : n \geq m\}$ *almost surely.*

Proof. We give an alternative proof. By (i) one has

$$\sup_{n\in\mathbf{N}} \int_\Omega \langle e^*, \frac{S_n}{n} \rangle dP < \infty$$

for all $e^* \in E$. Then by Kingmann ergodic theorem, for each $f^* \in E^*_+$, $\langle f^*, \frac{S_n}{n} \rangle$ converges a.s. and also in L^1 to the integrable function

$$\sup_{n\in\mathbf{N}} \frac{E^{\mathcal{I}} \langle f^*, S_n \rangle}{n} = \sup_{n\in\mathbf{N}} \langle f^*, \frac{E^{\mathcal{I}} S_n}{n} \rangle$$

which entails

$$\lim_{n\to\infty} \langle f^*, \int_A \frac{S_n}{n} dP \rangle = \lim_{n\to\infty} \int_A \langle f^*, \frac{S_n}{n} \rangle dP = \int_A \sup_{n\in\mathbf{N}} \langle f^*, \frac{E^{\mathcal{I}} S_n}{n} \rangle dP \quad \forall A \in \mathcal{F}$$

Since by (i) the sequence $(\frac{S_n}{n})_{n\in\mathbf{N}}$ is bounded in $L^1_E(\Omega, \mathcal{F}, P)$ and E is reflexive, there is a sequence $(Z_n)_{n\in\mathbf{N}}$ whose members are convex combinations of $(\frac{S_n}{n})$ and $Z_\infty \in L^1_E(\Omega, \mathcal{F}, P)$ such that

(7.2.1) $$\lim_{n\to\infty} Z_n(\omega) = Z_\infty(\omega) \quad a.s.$$

(see e.g. [11], Proposition 6.7), that is $(Z_n)_{n\in N}$ norm converges a.s. to Z_∞. Now we can finish the proof as in Theorem 7.1. ■

Remarks. Theorem 7.1 can be considered as a by product of a convergence result related to the Fatou lemma in Mathematical Economics. See ([12], Theorem 4.9). That is a surprise establishing the link between convergence in superadditive process in Probability and Fatou Lemma in Mathematical Economics. Theorem 7.1 is an extension of Kingman ergodic theorem to vector valued integrable superadditive sequences. In both Theorem 7.1 and its corollary, it is easily seen that the sequence $(\frac{S_n}{n})_{n\in\mathbf{N}}$ is Mazur-tight in the sense of [12]. The use of Mazur trick is decisive here.

8. Convex weakly compact valued superadditive process in $\mathcal{L}^1_{cwk(E)}(\Omega, \mathcal{F}, P)$

In this section we will develop some Komlós techniques in the convergence of convex weakly compact valued superadditive process via the Kingman ergodic theorem for superadditive process in $L^1_{\mathbf{R}}(\Omega, \mathcal{F}, P)$. We need to provide the definition of superadditive process in $\mathcal{L}^1_{cwk(E)}(\Omega, \mathcal{F}, P)$.

Definition 8.1. *A sequence $(S_m)_{m\in\mathbf{N}}$ in $\mathcal{L}^1_{cwk(E)}(\Omega, \mathcal{F}, P)$ is a superadditive process provided that for all $m, n \in \mathbf{N}$, for all $\omega \in \Omega$, we have*

$$S_m(\omega) + S_n(T^m\omega) \subset S_{m+n}(\omega).$$

Further, the superadditivity given in this definition is equivalent to

$$\delta^*(x^*, S_m(\omega)) + \delta^*(x^*, S_n(T^m\omega)) \leq \delta^*(x^*, S_{m+n}(\omega)) \quad \forall x^* \in E^*.$$

so that coming back to the definition of real valued integrable superadditive process, we see that, for each $x^* \in E^*$, the sequence $(\delta^*(x^*, S_m))_{m\in\mathbf{N}}$ is a superadditive integrable process in $L^1_{\mathbf{R}}(\Omega, \mathcal{F}, P)$. When equality is assumed to hold in the definition 8.1, $(S_m)_{m\in\mathbf{N}}$ is called *additive*. Compare with [28]. In the remainder of this paper, E is a separable Hilbert space. The following lemma is a useful and has its own interest.

Lemma 8.1. *Let E be a separable Hilbert space. Let $(X_n)_{n\in\mathbf{N}}$ be a bounded sequence in $\mathcal{L}^1_{cwk(E)}(\Omega, \mathcal{F}, P)$. Then there is a subsequence $(X_{\gamma(n)})_{n\in\mathbf{N}}$ of $(X_n)_{n\in\mathbf{N}}$ and X_∞ in $\mathcal{L}^1_{cwk(E)}(\Omega, \mathcal{F}, P)$ such that*

$$\lim_{n\to\infty} \delta^*(x, \frac{1}{n}\sum_{i=1}^{n} X_{\gamma(i)}(\omega)) = \delta^*(x, X_\infty(\omega)) \quad a.s. \quad \forall x \in \overline{B}_E.$$

$$\lim_{n\to\infty} d(x, \frac{1}{n}\sum_{i=1}^{n} X_{\gamma(i)}(\omega)) = d(x, X_\infty(\omega)) \quad a.s. \quad \forall x \in E.$$

Proof. Step 1 Let $(e_k)_{k\in\mathbf{N}}$ be a dense sequence in $\overline{B}_E$. For each fixed $k \in \mathbf{N}$ and $n \in \mathbf{N}$, pick an integrable selection $u_{k,n}$ of X_n such that

$$\langle e_k, u_{k,n}(\omega)\rangle = \delta^*(e_k, X_n(\omega)) \quad \forall \omega \in \Omega.$$

As the sequence $(|X_n|)_{n\in\mathbf{N}}$ is bounded in $L^1_{\mathbf{R}}(\Omega, \mathcal{F}, P)$ and the sequence $(u_{k,n})_{n\in\mathbf{N}}$ is bounded in $L^1_E(\Omega, \mathcal{F}, P)$ for each $k \in \mathbf{N}$, by applying Komlós theorem[2] in Hilbert space [23] and a diagonal procedure we find an application $\gamma : \mathbf{N} \to \mathbf{N}$, $\alpha \in L^1_{\mathbf{R}}(\Omega, \mathcal{F}, P)$, $u_{k,\infty} \in L^1_E(\Omega, \mathcal{F}, P)$ satisfying

$$\lim_{n\to\infty} \frac{1}{n}\sum_{i=1}^{n} |X_{\gamma(i)}|(\omega) = \alpha(\omega) \quad a.s. \tag{8.1.1}$$

and for each $k \in \mathbf{N}$

$$\lim_{n\to\infty} \frac{1}{n}\sum_{i=1}^{n} u_{k,\gamma(i)}(\omega) = u_{k,\infty}(\omega) \quad a.s. \tag{8.1.2}$$

with respect to the norm of E.
Step 2 Let us set

$$Y_n(\omega) := \frac{1}{n}\sum_{i=1}^{n} X_{\gamma(i)}(\omega) \quad \forall \omega \in \Omega$$

so that by (8.1.1) $\sup_{n\in\mathbf{N}} |Y_n| < \infty$ almost surely; and

$$X_\infty(\omega) = s\text{-}li\ Y_n(\omega) \quad \forall \omega \in \Omega$$

so that for every $k \in \mathbf{N}$

$$u_{k,\infty}(\omega) \in X_\infty(\omega) \quad a.s \quad \omega \in \Omega.$$

Then for every $k \in \mathbf{N}$ we have

$$\lim_{n\to\infty} \delta^*(e_k, Y_n(\omega)) = \lim_{n\to\infty} \frac{1}{n}\sum_{i=1}^{n} \langle e_k, u_{k,\gamma(i)}\rangle$$
$$= \langle e_k, u_{k,\infty}(\omega)\rangle \le \delta^*(e_k, X_\infty(\omega) \quad a.s.$$

[2] A more general version of Komlós theorem for B convex reflexive Banach space is available in [21].

and

$$\delta^*(e_k, X_\infty(\omega)) \le \lim_{n\to\infty} \delta^*(e_k, Y_n(\omega)) \quad a.s.$$

so that

$$\delta^*(e_k, X_\infty(\omega)) = \lim_{n\to\infty} \delta^*(e_k, Y_n(\omega)) \quad a.s.$$

Using a density argument as in the previous results in section 6 we deduce that

$$\text{(8.1.2)} \qquad \delta^*(x, X_\infty(\omega)) = \lim_{n\to\infty} \delta^*(x, Y_n(\omega)) \quad a.s. \quad \forall x \in E.$$

Finally X_∞ is measurable and satisfies the inequality

$$\int_\Omega |X_\infty| dP \le \int_\Omega \liminf_n |Y_n| dP \le \sup_n \int_\Omega |X_n| dP < \infty.$$

Step 3 Now we claim that

$$\lim_{n\to\infty} d(x, Y_n(\omega)) = d(x, X_\infty(\omega)) \quad a.s. \quad \forall x \in \overline{B}_E.$$

From (8.1.2) deduce that

$$\text{(8.1.3)} \qquad \liminf_n d(x, Y_n(\omega))$$

$$= \liminf_n \sup_{y\in\overline{B}_E} [\langle y, x\rangle - \delta^*(y, Y_n(\omega))]$$

$$\ge \sup_k \lim_n [\langle e_k, x\rangle - \delta^*(e_k^*, Y_n(\omega))]$$

$$= \sup_k [\langle e_k, x\rangle - \delta^*(e_k, X_\infty(\omega))]$$

$$= d(x, X_\infty(\omega)).$$

for all $x \in E$ and almost surely $\omega \in \Omega$. By definition of X_∞ we have

$$\text{(8.1.4)} \qquad \limsup_n d(x, Y_n(\omega)) \le d(x, X_\infty(\omega))$$

for all $x \in E$ and all $\omega \in \Omega$. So the claim follows from (8.1.3)-(8.1.4). ■

Now we provide our convergence theorem for convex weakly compact valued superadditive process in Hilbert spaces.

Theorem 8.1. *Assume that E is separable Hilbert space, T is a measurable transformation of Ω preserving P, $\mathcal{I}$ is the σ-algebra of invariant sets and $(S_n)_{n\in\mathbf{N}}$ is a superadditive process in $\mathcal{L}^1_{cwk(E)}(\Omega, \mathcal{F}, P)$ satisfying*

(i) $(\frac{S_n}{n})_{n\in\mathbf{N}}$ *is uniformly integrable.*
(ii) There is $L \in \mathcal{L}wk(E)$ *such that* $\frac{S_n}{n}(\omega) \subset L, \forall n \in \mathbf{N}, \forall \omega \in \Omega$.

Then there are a subsequence $(Z_{\gamma(n)})_{n\in\mathbf{N}}$ *of* $(\frac{S_n}{n})_{n\in\mathbf{N}}$ *and* $Z_\infty \in \mathcal{L}^1_{cwk(E)}(\Omega, \mathcal{F}, P)$ *satisfying the following properties:*

$$\lim_{n\to\infty} \delta^*(x, \frac{1}{n}\sum_{i=1}^{n} Z_{\gamma(i)}(\omega)) = \delta^*(x, Z_\infty(\omega)) \quad a.s. \quad \forall x \in \overline{B}_E.$$

$$\lim_{n\to\infty} d(x, \frac{1}{n}\sum_{i=1}^{n} Z_{\gamma(i)}(\omega)) = d(x, Z_\infty(\omega)) \quad a.s. \quad \forall x \in E.$$

Consequently

$$\int_A \delta^*(x, Z_\infty)dP = \lim_{n\to\infty}\int_A \delta^*(x, \frac{S_n}{n})dP = \int_A \sup_{n\in\mathbf{N}} \frac{\delta^*(x, E^{\mathcal{I}}S_n)}{n}dP$$

for each $x \in \overline{B}_E$ *and each* $A \in \mathcal{F}$ *with the localization property*

$$Z_\infty(\omega) \subset \overline{co}\, w\text{-}ls \frac{S_n(\omega)}{n} \quad a.s.$$

Proof. Let $x \in \overline{B}_E$. Then by our assumption and by virtue of Kingman ergodic theorem, the uniformly integrable real valued superadditive sequence $((\delta^*(x, \frac{S_n}{n}))_{n\in\mathbf{N}}$ converges a.s. and in L^1 to the integrable function

$$\gamma_x := \sup_{n\in\mathbf{N}} \frac{E^{\mathcal{I}}\delta^*(x, S_n)}{n} = \sup_{n\in\mathbf{N}} \frac{\delta^*(x, E^{\mathcal{I}}S_n)}{n}$$

which entails

$$\lim_{n\to\infty}\int_A \delta^*(x, \frac{S_n}{n})dP = \int_A \sup_{n\in\mathbf{N}} \frac{\delta^*(x, E^{\mathcal{I}}S_n)}{n}dP \tag{8.2.1}$$

for each $A \in \mathcal{F}$. Since $(\frac{S_n}{n})_{n\in\mathbf{N}}$ is bounded in $\mathcal{L}^1_{cwk(E)}(\Omega, \mathcal{F}, P)$, by Lemma 8.1 we find a subsequence $(Z_{\gamma(n)})_{n\in\mathbf{N}}$ of $(\frac{S_n}{n})_{n\in\mathbf{N}}$ and $Z_\infty \in \mathcal{L}^1_{cwk(E)}(\Omega, \mathcal{F}, P)$ satisfying

$$\lim_{n\to\infty} \delta^*(x, \frac{1}{n}\sum_{i=1}^{n} Z_{\gamma(i)}(\omega)) = \delta^*(x, Z_\infty(\omega)) \quad a.s. \quad \forall x \in \overline{B}_E. \tag{8.2.2}$$

$$\lim_{n\to\infty} d(x, \frac{1}{n}\sum_{i=1}^{n} Z_{\gamma(i)}(\omega)) = d(x, Z_\infty(\omega)) \quad a.s. \quad \forall x \in E. \tag{8.2.3}$$

From (8.2.1)-(8.2.2) and Lebesgue-Vitali theorem, it follows that

$$\lim_{n\to\infty}\int_A \delta^*(x, \frac{S_n(\omega)}{n})dP(\omega) = \lim_{n\to\infty}\int_A \delta^*(x, \frac{1}{n}\sum_{i=1}^{n} Z_{\gamma(i)}(\omega))dP(\omega)$$

$$= \int_A \delta^*(x, Z_\infty(\omega))dP(\omega) \tag{8.2.4}$$

for every $x \in E$ and for every $A \in \mathcal{F}$. Now we claim that

$$Z_\infty(\omega) \subset \overline{co}\, w\text{-}ls\frac{S_n(\omega)}{n} \quad a.s. \tag{8.2.5}$$

As $\frac{S_n(\omega)}{n}$ is included in $L \in \mathcal{L}wk(E)$, the $\mathcal{L}wk(E)$-valued mapping $\omega \to \overline{co}\, w\text{-}ls\frac{S_n(\omega)}{n}$ is measurable, see e.g. ([12], Theorem 5.4). Assume that (8.2.5) does not hold. Using Lemma III. 34 in [16] provides $x \in E$ and a $\mathcal{F}$-measurable set A with $P(A) > 0$ such that

$$\delta^*(x, Z_\infty(\omega)) > \delta^*(x, \overline{co}\, w\text{-}ls\frac{S_n(\omega)}{n})$$

for all $\omega \in A$. For each $n \in \mathbf{N}$ pick a maximal integrable selection z_n of $\frac{S_n}{n}$ associated with x, i.e.

$$\langle x, z_n(\omega)\rangle = \delta^*(x, \frac{S_n(\omega)}{n}) \quad \forall \omega \in \Omega.$$

Then it is obvious that the sequence $(z_n)_{n\in\mathbf{N}}$ is relatively sequentially compact in $L^1_E(\Omega, \mathcal{F}, P)$, see e.g. ([13], Theorem 6.2.5 or Theorem 6.4.13). So we may assume that $(z_n)_{n\in\mathbf{N}}$ $\sigma(L^1_E, L^\infty_{E'})$ converges to a function $z_\infty \in L^1_E(\Omega, \mathcal{F}, P)$ with the localization property ([13], Proposition 6.5.3)

$$z_\infty(\omega) \in \overline{co}\, w\text{-}ls\, z_n(\omega) \subset \overline{co}\, w\text{-}ls\frac{S_n(\omega)}{n} \quad a.s.$$

As $z_n \to z_\infty$ weakly in $L^1_E(\Omega, \mathcal{F}, P)$ we get

$$\lim_{n\to\infty}\int_A \langle x, z_n\rangle dP = \int_A \langle x, z_\infty\rangle dP.$$

By combining with (8.2.4) it follows that

$$\int_A \langle x, z_\infty(\omega)\rangle dP(\omega) = \int_A \delta^*(x, Z_\infty(\omega))dP(\omega).$$

As $z_\infty(\omega) \in \overline{co}\, w\text{-}ls \frac{S_n(\omega)}{n}$ a.s. on A, by integrating on A we deduce that

$$\int_A \delta^*(x, \overline{co}\, w\text{-}ls \frac{S_n(\omega)}{n})dP(\omega) \geq \int_A \delta^*(x, Z_\infty(\omega))dP(\omega)$$

that contradicts the inequality

$$\int_A \delta^*(x, \overline{co}\, w\text{-}ls \frac{S_n(\omega)}{n})dP(\omega) < \int_A \delta^*(x, Z_\infty(\omega))dP(\omega).$$

■

Remarks. 1) Lemma 8.1 and Theorem 8.2 hold true when E is separable B convex reflexive Banach space. If $(S_n)_{n\in\infty}$ is additive, Theorem 8.2 needs a special treatment. Compare this variant with [28] and Theorems 6.1–6.6. For shortness we omit the details.

2) In most results presented here we need to deal with convex weakly-weakly star compact valued random sets (alias measurable multifunctions), apart from the case of epiconvergence theorem for normal integrands in Proposition 5.1 involving the support function of an epigraph multifunction and the support function of the conditional expectation of this epigraph. It seems difficult to obtain the above convergence results in the case when the random sets are unbounded.

3) It would be interesting to obtain the analogues of the aforementioned results for subadditive and superstationary convex weakly-weakly star compact valued process satisfying some tightness conditions (see e.g. [12, 14]).

References

1. Abid, M.: Un théoreme ergodique pour les processus sous-additifs et sur-stationnaires. C.R. Acad. Sci. Paris, Serie A **287**, 149–152 (1978)
2. Akhiat, F., Castaing, C., Ezzaki, F.: Some various convergence results for multivalued martingales. In: Kusuoka, S., Maruyama, T. (eds.) Adv. Math. Econ. **13**, 1–33 (2010)
3. Beer, G.: Support and distance functionals for convex sets. Numer. Funct. Anal. Optim. **10**(1–2), 15–36 (1989)
4. Beer, G.: The slice topology: a viable alternative to Mosco convergence in nonreflexive spaces. Nonlinear Anal. **19**(3), 271–290 (1992)
5. Castaing, C.: Compacité et inf-equicontinuity dans certains espaces de Köthe-Orlicz. Séminaire d'Analyse Convexe, Montpellier, Exposé No 6 (1979)

6. Castaing, C.: Some various convergence results for normal integrands. Adv. Math. Econ. **15**, 1–26 (2011)
7. Castaing, C., Ezzaki, F.: Variational inequalities for integrand martingales and additive random sequences. Séminaire d'Analyse Convexe, Montpellier, Exposé No 1 (1992) and Acta Mathematica Vietnamica **18**(1), 137–171 (1993)
8. Castaing, C., Ezzaki, F., Hess, C.: Convergence for conditional expectation for unbounded closed convex random sets. Studia Mathematica **124**(2), 133–148 (1997)
9. Castaing, C., Ezzaki, F., Lavie, M., Saadoune, M.: Weak star convergence of martingales in a dual space. In: Proceedings of the 9-th Edition of the International Conference on Function Spaces, Krakow, Poland; Banach Center Publications, Vol 92. Institute of Mathematics, Polish Academy of Sciences, Warsawa (2011)
10. Castaing, C., Ezzaki, F., Tahri, K.: Convergences of multivalued pramarts. Journal of Nonlinear and Convex Analysis **11**(2), 243–266 (2010)
11. Castaing, C., Guessous, M.: Convergences in $L^1_X(\mu)$. Adv. Math. Econ. **1**, 17–37 (1999)
12. Castaing, C., Hess, Ch., Saadoune, M.: Tightness conditions and integrability of the sequential weak upper limit of a sequence of multifunctions. Adv. Math. Econ. **11**, 11–44 (2008)
13. Castaing, C., Raynaud de Fitte, P., Valadier, M.: Young Measures on Topological Spaces. With Applications in Control Theory and Probability Theory. Kluwer Academic Publishers, Dordrecht (2004)
14. Castaing, C., Saadoune, M.: Dunford-Pettis types theorem and convergences in set-valued integration. Journal of Nonlinear and Convex Analysis **1**(1), 37–71 (2000)
15. Castaing, C., Saadoune, M.: Convergences in a dual space with applications to Fatou lemma. In: Kusuoka, S., Maruyama, T. (eds.) Adv. Math. Econ. **12**, pp. 23–69 (2009)
16. Castaing, C., Valadier, M.: Convex Analysis and Measurable Multifunctions, Lecture Notes in Math., Vol. 580. Springer-Verlag, Berlin and New York (1977)
17. Choirat, C., Hess, C., Seri, R.: A functional version of the Birkhoff ergodic theorem for a normal integrand: a variational approach. The Annals of Probability **31**, 63–92 (2003)
18. Dudley, R. M.: Real Analysis And Probability, Chapman-Hall, Mathematics Series. Wadsworth, Inc. (1989)
19. Fitzpatrick, S., Lewis, A.S.: Weak-star convergence of convex sets. Journal of Convex Analysis **13**(3 + 4), 711–719 (2006)
20. Hiai, F., Umegaki, H.: Integrals, conditional expectations and martingales of multivalued functions. J. Multi. Anal. **7**, 149–182 (1977)

21. Garling, D. J. H.: Subsequence principles for vector-valued random variables. Math. Proc. Cambridge Philos. Soc. **86**, 301–311 (1979)
22. Ghoussoub, N., Steele, M.: Vector valued subadditive process and applications in probability. Ann. Probability **8**(1), 83–95 (1990)
23. Guessous, M.: An elementary proof of Komlos-Revesz theorem in Hilbert space. Journal of Convex Analysis **4**, 321–332 (1997)
24. Jalby, V.: Semi-continuité, convergence et approximation des applications vectorielles. Loi des grands nombres, Université Montpellier II, Laboratoire d'Analyse Convexe, 34095 Montpellier Cedex 05, France, Janvier (1992)
25. Korf, L. A., Wets, R. J. B.: Random l.s.c functions: an ergodic theorem. Mathematics of Operations Research **26**(2), 421–445 (2001)
26. Krengel, U.: Ergodic Theorems. De Gruyter Studies in Mathematics 6, Berlin, New York (1985)
27. Krupa, G.: Ergodic theorems for subadditive superstationary families of random sets with values in Banach spaces. Studia Mathematica **131**(3), (1998)
28. Mourier, E.: Eléments aléatoires dans un espace de Banach. Ann. Ins. H. Poincare, Sect. B **13**, 161–244 (1953)
29. Neveu, N.: Martingales Etemps discret, Masson et Cie, Editeur (1972)
30. Peressini, A. L.: Ordered Topological Vector Spaces, Harper's Series in Modern Mathematics. Harper and Row Publ., New-York (1967)
31. Schaefer, H.H.: Banach Lattices and Positive Operators, Die Grundlehren der Mathematischen Wissenschaften in Einzeldarstellungen, **215**. Springer-Verlag, Berlin
32. Schurger, K.: Ergodic theorems for subadditive superstationary families of convex random sets. Z. Wahrsch. verw. Gebiete **62**, 125–135 (1983)
33. Valadier, M.: On conditional expectation of random sets. Annali Math. Pura Appl. **126**, 81–91 (1980)
34. Valadier, M.: Conditional expectation and ergodic theorem for a positive integrand. J. Nonlinear Convex Anal. **1**, 233–244 (2002)

Adv. Math. Econ. 16, 39–51 (2012)

Advances in
MATHEMATICAL
ECONOMICS

Elementary form and proof of the Frobenius theorem for economists

Yuhki Hosoya

Graduate School of Economics, Keio University
(e-mail: ukki@jcom.home.ne.jp)

Received: May 11, 2011
Revised: May 18, 2011

JEL classification: D11

Mathematics Subject Classification (2010): 53D45, 91B16

Abstract. This note presents an elementary form of the Frobenius theorem and provides its proof. It studies a condition that is related to the symmetry of the Antonelli matrix and the classical integrability condition.

Key words: Frobenius theorem, integrability, Antonelli matrix

1. Introduction

In consumer theory, economists often need to use the **Frobenius theorem**. This note aims to present an elementary form of this theorem and to provide its proof. The merits of this note are explained below:

i. Readers need to have only basic mathematical knowledge to understand this note.
   Most descriptions of the Frobenius theorem[1] are very difficult to comprehend because they require deep knowledge of either differential forms or vector fields. Since many economists do not have this knowledge,

[1] See, for example, Matsushima [8], Kosinski [7], Hicks [5], Auslander and MacKenzie [2], and Sternberg [11].

they cannot check the proof of the Frobenius theorem. In contrast, to understand our claim and to check our proof, readers only need to know some elementary facts on general topology, linear algebra, implicit and inverse function theorem, and ordinary differential equations (ODEs).

ii. This note clarifies that this theorem only requires C^1-class smoothness on the system.
Most claims of the Frobenius theorem require C^∞-class smoothness on the system, and in fact, some of their proofs need at least C^2-class smoothness.[2] In contrast, our proof relaxes this condition by using some facts on ODEs. These facts include the existence and uniqueness of the solution of ODE, differentiability of the solution with regard to the parameters, and commutativity of second-order derivatives of the solution with regard to the parameters and times, that is, the equation $\partial_t \partial_y x(t; y) = \partial_y \partial_t x(t; y)$.[3] Using these facts, we confirm that this theorem still holds under C^1-class smoothness.[4]
Furthermore, our proof reveals that this theorem can no longer provide the necessary and sufficient condition of complete integrability when the system does not have C^∞-class smoothness. Debreu [3] makes a mistake in this regard; he indicates this and provides a counterexample in its corrigendum (Debreu [4]). Our proof clarifies this aspect, and thus, helps the readers avoid such mistakes.

iii. In our theorem, both the symmetry on the **Antonelli matrix** and the classical integrability condition appear naturally.
The concept of the Antonelli matrix is very old and popular in integrability theory, and many works consider the property of this matrix. Furthermore, it is well known that this matrix corresponds much with the Slutsky matrix. Most articles that discuss the demand theory mention the symmetry of the Antonelli matrix. Antonelli [1] was the first articles to mention the elements of this matrix. Samuelson [10] discussed the symmetry and other properties of the Antonelli matrix, and characterized the condition of integrability by some properties of this matrix. Hurwicz [6] is an interesting survey in this context.
Hence, the Antonelli matrix is an important tool that should be studied in order to understand the classical demand theory. The Antonelli matrix

[2] The proof of Matsushima [8] is a typical example of this. His proof uses the fact $d \circ f^* = f^* \circ d$, where f^* is the pull-back operator defined by f and d is the exterior derivative operator. Since $d \circ f^*$ is well-defined only if the second-order derivatives of f exist, his proof needs at least C^2-class smoothness.

[3] To verify these facts, see Ch.4 of Pontryagin [9].

[4] As far as we know, there is no research that proves the Frobenius theorem under C^1-smoothness.

naturally appears in our theorem,[5] and the readers can easily relate our theorem to the many classical arguments in consumer theory.
Moreover, the classical integrability condition appears frequently in the context of integrability theory. Debreu [3] introduced this condition and stated that it corresponds with the condition of Lie bracket. This condition is equivalent to the symmetry of the Antonelli matrix whenever the n-th coordinate of the inverse demand function is always 1. Hence, readers can easily understand this condition from our theorem.

In sect. 2, we rewrite the claim of the Frobenius theorem in the elementary language of mathematics. Section 3 provides the proof.

2. Main Theorem

2.1. Our Theorem

Fix $n, m \in \mathbb{N}$ and suppose that $n > m \geq 1$. Let $A \subset \mathbb{R}^n$ be open and choose any $g_1, ..., g_m : A \to \mathbb{R}^n$ such that $\dim(\mathrm{span}\{g_1(x), ..., g_m(x)\}) = m$ for any $x \in A$. Fix any $x \in A$. Then, we state that $\{g_1, ..., g_m\}$ satisfies[6]

condition (A-q) at x if all $g_1, ..., g_m$ are C^{q-1}-class at x and $\exists U \subset A$ s.t. $x \in U$ and U is open, $\exists u : U \to \mathbb{R}^m$ s.t. u is C^q-class and $\mathrm{span}\{Du^1, ..., Du^m\} = \mathrm{span}\{g_1, ..., g_m\}$,

condition (B-q) at x if all $g_1, ..., g_m$ are C^q-class at x and $\exists U \subset A$ s.t. $x \in U$ and U is open, $\exists h_{1,1}, ..., h_{m,m} : U \to \mathbb{R}^n$ s.t. $\forall i$, $Dg_i^T - Dg_i = \sum_{j=1}^m [h_{i,j} g_j^T - g_j h_{i,j}^T]$,

condition (C-q) at x if all $g_1, ..., g_m$ are C^q-class at x and $\exists U \subset A$ s.t. $x \in U$ and U is open, $\exists \bar{g}_1, ..., \bar{g}_m : U \to \mathbb{R}^n$ s.t. $\exists I = \{i_1, ..., i_m\}$ s.t. $\bar{g}_j^{i_j} = 1$ and $\bar{g}_j^{i_k} = 0$ if $j \neq k$, $\mathrm{span}\{g_1, ..., g_m\} = \mathrm{span}\{\bar{g}_1, ..., \bar{g}_m\}$ on U and $\partial_k \bar{g}_i^j - \partial_j \bar{g}_i^k = \sum_{\ell=1}^m [\partial_{i_\ell} \bar{g}_i^j \bar{g}_\ell^k - \partial_{i_\ell} \bar{g}_i^k \bar{g}_\ell^j]$ for any $i \leq m$ and $j, k \notin I$.

Our main theorem is as follows.

Theorem 1. *Suppose $q \geq 1$. Then,*

1) condition (B-q) *at x implies condition* (C-q) *at x,*
2) condition (C-q) *at x implies condition* (A-q) *at x,*[7]
3) condition (A-$(q+1)$) *at x implies condition* (B-q) *at x.*

The next corollary clearly holds.

[5] See 2) of Theorem 1 in sect. 2.
[6] In this note, x^i denotes the i-th coordinate of vector x.
[7] The term f being C^0-class means that f is continuous.

Corollary 1. *The condition* (A-∞) *at* x, (B-∞) *at* x, *and* (C-∞) *at* x *are mutually equivalent.*

We should mention that if $q < \infty$, then both conditions (B-q) and (C-q) are not the sufficient condition of (A-($q+1$)). Debreu [4] presents a counterexample. He constructs the following g_1:

$$g_1(x, y) = \begin{cases} \left(\frac{y^2}{\sqrt{1+y^4}}, \frac{1}{\sqrt{1+y^4}}\right) & (\text{if } y \geq 0) \\ (0, 1) & (\text{otherwise}) \end{cases}$$

and claims that there is no C^2-class real-valued function u such that $Du(x, y) = \lambda(x, y)g_1(x, y)$ for any (x, y) around $(0, 0)$. Meanwhile, it can easily be verified that if $n = 2, m = 1$, and g_1 is C^1-class, then $\{g_1\}$ satisfies condition (B-1). Hence, in this case, condition (B-1) at $(0, 0)$ holds and condition (A-2) at $(0, 0)$ does not hold.

2.2. Relationships between Our Theorem and the Usual Frobenius Theorem

We should clarify that this theorem corresponds with the usual Frobenius theorem.[8] We assume that every function and manifold in this subsection has C^∞-class smoothness.

Let M be an n-dimensional manifold and D be a function defined on M such that for any $x \in M$, $D(x)$ is an $(n-m)$-dimensional subspace of the tangent space $T_x(M)$ of M at x. A collection of 1-forms $\omega_1, \ldots, \omega_m$ is said to be the local equations of D around x if there is some neighborhood U of x such that $D(y) = \cap_{i=1}^m \mathrm{Ker}\,\omega_i(y)$ for any $y \in U$. Suppose that D has at least one local equations. Now, D is said to be completely integrable around x if there are m real-valued functions $f_1, \ldots, f_m$ defined on some neighborhood of x such that the collection $df_1, \ldots, df_m$ comprises local equations of D around x, where df_i denotes the exterior derivative of f_i.

Theorem 2 (Frobenius). *D is completely integrable around x if and only if for any local equations $\omega_1, \ldots, \omega_m$, there exists a collection of 1-forms $\tilde{\omega}_{1,1}, \ldots, \tilde{\omega}_{m,m}$ such that for any $i \in \{1, \ldots, m\}$,*

$$d\omega_i = \sum_{j=1}^m \tilde{\omega}_{i,j} \wedge \omega_j,$$

[8] This form of the Frobenius theorem is obtained by Matsushima [8]. Although some forms of the Frobenius theorem appear to be different from this one, all of them are essentially equivalent to this form.

where $d\omega_i$ denotes the exterior derivative of ω_i, and $\tilde{\omega}_{i,j} \wedge \omega_j$ denotes the exterior product of $\tilde{\omega}_{i,j}$ and ω_j. (We call this condition the integrability condition.)

We show that the following claim holds:

Claim: Corollary 1 ensures that the Frobenius theorem holds.

First, this theorem claims the equivalence of the two properties *around* x. Hence, the claim of this theorem relates only to the local property, and thus, we can identify M with an open set $A \subset \mathbb{R}^n$. Then, it is well known that the tangent space of A at x, denoted by $T_x(A)$, is equal to $\mathbb{R}^n$, and therefore, the dual space of $T_x(A)$ is also equal to $\mathbb{R}^n$. Recall that the 1-form ω of M is a function on M such that $\omega(x)$ is an element of the dual space of $T_x(M)$. Thus, if $M = A$, the 1-form ω is simply a function $g : A \to \mathbb{R}^n$. Then, we can check that the exterior product $\omega_1 \wedge \omega_2$ corresponds to $g_1 g_2^T - g_2 g_1^T$. Furthermore, we can verify that the exterior derivative $d\omega$ corresponds to $Dg^T - Dg$. Finally, the exterior derivative of a real-valued function simply corresponds to the Jacobian matrix.

Under the above translation, we can check the following proposition.

Proposition 1. *1) $g_1, ..., g_m : A \to \mathbb{R}^n$ are local equations of D if and only if span$\{g_1(x), ..., g_m(x)\}^\perp = D(x)$ for any $x \in A$.*

2) Suppose $g_1, ..., g_m$ are local equations of D. Then, D is completely integrable around x if and only if condition (A-∞) *at x holds.*

3) Suppose $g_1, ..., g_m$ are local equations of D. Then, the integrability condition of the Frobenius theorem is equivalent to condition (B-∞) *at x.*

From 2) and 3) of Proposition 1, we have that the claim holds.

2.3. Application to the Classical Integrability Theory

Consider an open set U and a real-valued function $u : U \to \mathbb{R}$. u is interpreted as a utility function. A function $g : U \to \mathbb{R}$ is called an inverse demand function of u if and only if for any $x \in U$, x is a solution of the following maximization problem:

$$\begin{aligned} &\max_{y \in U} && u(y) \\ &\text{subject to} && g(x) \cdot y \le g(x) \cdot x. \end{aligned}$$

Then, it is well known that if g is an inverse demand function of u, $Du(x) = \lambda(x) g(x)$ for any x.[9]

[9] This condition is called the first-order condition.

Now, suppose that $m = 1$ and $\{g_1\}$ satisfies condition (A-q) at x. Then, g_1 is C^{q-1}-class, and there exists $u : U \to \mathbb{R}$ such that u is C^q-class and $\text{span}\{Du\} = \text{span}\{g_1\}$. Hence, there exists $\lambda : U \to \mathbb{R}$ such that λ is C^{q-1}-class and $Du = \lambda g_1$. Hence, the first-order condition is satisfied, and thus, u is a candidate of a utility function such that g_1 is an inverse demand function of u. We call this property the *local integrability of* g_1 around x.

Next, suppose again that $m = 1$ and $\{g_1\}$ satisfies condition (C-q) at x. Choose the corresponding $\bar{g}_1 : U \to \mathbb{R}^n$, and suppose that $i_1 = n$. Then, $\bar{g}_1(y) = (\bar{g}_1^1(y), ..., \bar{g}_1^{n-1}(y), 1)$ for any $y \in U$ and $\text{span}\{g_1\} = \text{span}\{\bar{g}_1\}$, which implies that $\bar{g}_1(y) = \frac{1}{g_1^n(y)} g_1(y)$ for any $y \in U$. By condition (C-q) at x,

$$\partial_k \bar{g}_1^j - \partial_j \bar{g}_1^k = \partial_n \bar{g}_1^j \bar{g}_1^k - \partial_n \bar{g}_1^k \bar{g}_1^j,$$

which is equivalent to

$$\partial_k \bar{g}_1^j - \partial_n \bar{g}_1^j \bar{g}_1^k = \partial_j \bar{g}_1^k - \partial_n \bar{g}_1^k \bar{g}_1^j.$$

The $(n-1) \times (n-1)$ matrix $A = (a^{i,j})$ is called the **Antonelli matrix** of $\bar{g}_1$ if $a^{i,j} = \partial_j \bar{g}_1^i - \partial_n \bar{g}_1^i \bar{g}_1^j$. Then, the above equation claims that the Antonelli matrix of $\bar{g}_1$ is symmetric. Thus, we can see that 2) of our Theorem 1 indicates that the symmetry of the Antonelli matrix yields the existence of u such that $Du = \lambda \bar{g}_1 = \frac{\lambda}{g_1^n} g_1$ around x.

Third, suppose that $g_1 : A \to \mathbb{R}^n$ is C^q-class. Consider the following equation:

$$g_1^i(\partial_j g_1^k - \partial_k g_1^j) + g_1^j(\partial_k g_1^i - \partial_i g_1^k) + g_1^k(\partial_i g_1^j - \partial_j g_1^i) \equiv 0. \qquad (1)$$

This equation is also called the *integrability condition*. An important property of this equation is the invariance of normalization. That is, if $\sigma : A \to \mathbb{R} \setminus \{0\}$ is C^q-class and $h : x \mapsto \sigma(x) g_1(x)$, then h satisfies the above equation if and only if g_1 satisfies it. Now, suppose that $\sigma(x) = \frac{1}{g_1^n(x)}$ for all x. Then, $h^n(x) \equiv 1$, and we can easily check that the integrability condition is equivalent to the symmetry of the Antonelli matrix of h. Hence, $\{g_1\}$ satisfies condition (C-q) if and only if g_1 satisfies the integrability condition. As we have seen, this property ensures the existence of u such that $Du = \lambda g_1$ around x. Hence, we have verified the following classical result.

Proposition 2. *Both the symmetry of the Antonelli matrix of* $\bar{g}_1$ *and Equation (1) ensure the local integrability of* g_1.

3. Proof

3.1. Lemma

Recall that $n, m \in \mathbb{N}, n > m \geq 1$, the set $A \subset \mathbb{R}^n$ is open, and $g_1, ..., g_m : A \to \mathbb{R}^n$ satisfies $\dim(\mathrm{span}\{g_1(x), ..., g_m(x)\}) = m$ for any $x \in A$. First, we present a lemma.

Lemma 1. *Suppose $U_1 \subset A$ is open and $x \in U_1$. Choose any $\bar{g}_1, ..., \bar{g}_m : U_1 \to \mathbb{R}^n$ such that all $\bar{g}_i$ are C^q-class and $\mathrm{span}\{\bar{g}_1, ..., \bar{g}_m\} = \mathrm{span}\{g_1, ..., g_m\}$ on U_1. Then, $\{g_1, ..., g_m\}$ satisfies condition (B-q) at x if and only if $\{\bar{g}_1, ..., \bar{g}_m\}$ satisfies condition (B-q) at x.*

Remark. Suppose that $m = 1$. Then, the requirement $\mathrm{span}\{\bar{g}_1\} = \mathrm{span}\{g_1\}$ means that there exists some $\lambda : U_1 \to \mathbb{R} \setminus \{0\}$ such that $g_1 = \lambda \bar{g}_1$ on U_1. Hence, this lemma shows that condition (B-q) is independent from any change in the normalization of g_1.

Proof. We provide the proof of the "only if" part. The proof of the "if" part can be shown using the symmetric argument.

By assumption, there exists $(a_{i,j}), (b_{i,j})$ such that all $a_{i,j}, b_{i,j}$ are C^q-class, $\bar{g}_i = \sum_{j=1}^m a_{i,j} g_j$, and $g_i = \sum_{j=1}^m b_{i,j} \bar{g}_j$. Moreover, since $\{g_1, ..., g_m\}$ satisfies condition (B-q), we can choose the corresponding $U \subset U_1$ and $h_{1,1}, ..., h_{m,m}$. Define

$$\bar{h}_{i,j}^k = \sum_{\ell=1}^m [-\partial_k a_{i,\ell} + \sum_{p=1}^m a_{i,p} h_{p,\ell}^k] b_{\ell,j}.$$

By simple computation, we can verify that $\bar{h}_{i,j}$ satisfies our desired condition. Hence, $\{\bar{g}_1, ..., \bar{g}_m\}$ satisfies condition (B-q) at x. ■

3.2. Proof of 1)

Suppose that M is the $m \times n$ matrix such that the i-th row of M is g_i^T. Since $\dim(\mathrm{span}\{g_1(x), ..., g_m(x)\}) = m$, there exists $i_1, ..., i_m \in \{1, ..., n\}$ such that

$$N = \begin{pmatrix} g_1^{i_1} & \cdots & g_1^{i_m} \\ \vdots & \ddots & \vdots \\ g_m^{i_1} & \cdots & g_m^{i_m} \end{pmatrix}$$

is regular at x. Let $U_1 = \{y \in A | N(y) \text{ is regular}\}$ and define $\bar{M} = N^{-1}M$; further, let $\bar{g}_i$ be the i-th column of $\bar{M}^T$ on U_1. Clearly, $x \in U_1$ and U_1 is open.

Clearly, $\mathrm{Ker}\, M = \mathrm{Ker}\, \bar{M}$, and thus, $\mathrm{span}\{g_1, ..., g_m\}^{\perp} = \mathrm{span}\{\bar{g}_1, ..., \bar{g}_m\}^{\perp}$; this implies that $\mathrm{span}\{g_1, ..., g_m\} = \mathrm{span}\{\bar{g}_1, ..., \bar{g}_m\}$. By Lemma 1, $\{\bar{g}_1, ..., \bar{g}_m\}$ satisfies condition (B-q) at x, and thus, there exists the corresponding $U \subset U_1$ and $(h_{i,j})_{i,j}$. Hence, our goal is to show that for any $i \le m$ and $j, k \notin I \equiv \{i_1, ..., i_m\}$,

$$\partial_k \bar{g}_i^j - \partial_j \bar{g}_i^k = \sum_{\ell=1}^{m} [\partial_{i_\ell} \bar{g}_i^j \bar{g}_\ell^k - \partial_{i_\ell} \bar{g}_i^k \bar{g}_\ell^j],$$

on U.

By condition (B-q),

$$\partial_k \bar{g}_i^\ell - \partial_\ell \bar{g}_i^k = \sum_{j=1}^{m} [h_{i,j}^k \bar{g}_j^\ell - h_{i,j}^\ell \bar{g}_j^k].$$

If $k = i_a \in I$, the left-hand side is $\partial_k \bar{g}_i^\ell$ and the right-hand side is $\sum_{j=1}^m h_{i,j}^k \bar{g}_j^\ell - h_{i,a}^\ell$. Hence,

$$\partial_k \bar{g}_i^\ell = \sum_{j=1}^{m} h_{i,j}^k \bar{g}_j^\ell - h_{i,a}^\ell$$

for any $k \in I$ and $\ell \le n$. In addition, if $\ell = i_b \in I$, then the left-hand side is 0 and the right-hand side is $h_{i,b}^k - h_{i,a}^\ell$. Thus,

$$h_{i,b}^k = h_{i,a}^\ell$$

for any $k, \ell \in I$.

Then, for any $j, k \notin I$,

$$\begin{aligned}
\sum_{\ell=1}^{m} [\partial_{i_\ell} \bar{g}_i^j \bar{g}_\ell^k - \partial_{i_\ell} \bar{g}_i^k \bar{g}_\ell^j] &= \sum_{\ell=1}^{m} \Big[\sum_{p=1}^{m} h_{i,p}^{i_\ell} \bar{g}_p^j \bar{g}_\ell^k - h_{i,\ell}^j \bar{g}_\ell^k \\
&\quad - \sum_{p=1}^{m} h_{i,p}^{i_\ell} \bar{g}_p^k \bar{g}_\ell^j + h_{i,\ell}^k \bar{g}_\ell^j \Big] \\
&= \sum_{\ell=1}^{m} (h_{i,\ell}^k \bar{g}_\ell^j - h_{i,\ell}^j \bar{g}_\ell^k) \\
&= \partial_k \bar{g}_i^j - \partial_j \bar{g}_i^k,
\end{aligned}$$

which completes the proof of 1). ■

3.3. Proof of 2)

Suppose that $\{g_1, ..., g_m\}$ satisfies condition (C-q) at x, and take the corresponding $\{\bar{g}_1, ..., \bar{g}_m\}, I = \{i_1, ..., i_m\}$ and U. Without loss of generality, we assume that $i_a = a$ for any $a \in \{1, ..., m\}$.

Define $x_1 = (x^1, ..., x^m)^T$ and $x_2 = (x^{m+1}, ..., x^n)^T$. For any $t \in \mathbb{R}$, $y \in \mathbb{R}^m$, and $z \in \mathbb{R}^{n-m}$ such that $(x_1 + y, x_2 + tz) \in U$, define

$$\Phi_i(t, y, z) = -\bar{g}_i(x_1 + y, x_2 + tz),$$

and consider the following initial value problem:

$$\begin{aligned} \dot{\phi}^i &= \textstyle\sum_{j=1}^{n-m} \Phi_i^{m+j}(t, \phi, z)z^j, \\ \phi(0) &= y. \end{aligned} \qquad (2)$$

Then, there exist $\varepsilon_1 > 0$, $U_1 \subset U - x$, and $\phi :]-\varepsilon_1, \varepsilon_1[\times U_1 \to \mathbb{R}^m$ such that $0 \in U_1$, U_1 is open and $\phi(\cdot; y, z)$ is a C^q-class solution of this problem.[10] Since $\phi(0; y, z) = y$ for any $(y, z) \in U_1$, we have that $D_y\phi(0; y, z)$ is the identity matrix. Hence, there exist $\varepsilon_2 > 0$ and open $U_2 \subset U_1$ such that $0 \in U_2$, and for any $t \in]-\varepsilon_2, \varepsilon_2[$ and $(y, z) \in U_2$, $D_y\phi(t; y, z)$ is regular.

Let $s = \frac{\varepsilon_2}{2}$, $U_3 = \{(y, z) | (y, s^{-1}z) \in U_2\}$, and define

$$\psi(t; y, z) = \phi(st; y, s^{-1}z)$$

for any $(t, y, z) \in]-2, 2[\times U_3$. Then, U_3 is open, $\psi(0; y, z) = y$, $D_y\psi(t; y, z)$ is always regular, and

$$\dot{\psi}^i(t; y, z) = \sum_{j=1}^{n-m} \Phi_i^{m+j}(t, \psi(t; y, z), z)z^j,$$

and thus, ψ is a solution of (2) on $]-2, 2[\times U_3$.

Define $F(y, z, u) = y - \psi(1; u, z)$. Then, D_uF is regular whenever $(u, z) \in U_3$. Moreover, we have $\psi(1; 0, 0) = 0$, and thus, $F(0, 0, 0) = 0$. Hence, by the implicit function theorem, there exist open $U_4 \subset U_3$, $V_4 \subset \mathbb{R}^m$, and $\hat{u} : U_4 \to V_4$ such that $\hat{u}$ is a C^q-class function, $(0, 0, 0) \in U_4 \times V_4$, $\hat{u}(0, 0) = 0$, and for any $(y, z, u) \in U_4 \times V_4$,

$$F(y, z, u) = 0 \Leftrightarrow u = \hat{u}(y, z).$$

Define $\bar{u}(y, z) = (\hat{u}(y, z), z)$ for any $(y, z) \in U_4$. Since $F(y, 0, u) = y - u$, $D_y\hat{u}(0, 0)$ is regular, and thus, $D\bar{u}(0, 0)$ is also regular. Hence, by the inverse function theorem, there exist open $U_5 \subset U_4$ and $U_6 \subset \mathbb{R}^n$ such that $(0, 0) \in$

[10] Note that ε_1 can be taken uniformly on U_1 whenever U_1 is sufficiently small.

$U_5 \cap U_6$, $\bar{u}$ is a C^q-class bijection from U_5 onto U_6, and $\bar{u}^{-1}$ is also C^q-class. Without loss of generality, we assume that $U_6 \subset U_4$ and $U_6 = V_6 \times W_6$ for some open $V_6 \subset \mathbb{R}^m$ and $W_6 \subset \mathbb{R}^{n-m}$.

Fix any y and choose any (t, z) such that $(y, z) \in U_4$ and $t \in]-2, 2[$. Then, we define

$$\Psi_y(t, z) = (x_1 + \psi(t; y, z), x_2 + tz),$$

$$A_y(t, z) = \begin{pmatrix} \bar{g}_1^1(\Psi_y(t, z)) & \cdots & \bar{g}_1^n(\Psi_y(t, z)) \\ \vdots & & \vdots \\ \bar{g}_m^1(\Psi_y(t, z)) & \cdots & \bar{g}_m^n(\Psi_y(t, z)) \end{pmatrix} D_z \Psi_y(t, z),$$

that is,

$$A_y^{i,j}(t, z) = \bar{g}_i^T(\Psi_y(t, z)) D_{z^j} \Psi_y(t, z) = \partial_{z^j} \psi^i(t; y, z) + t\bar{g}_i^{m+j}(\Psi_y(t, z)).$$

Clearly, $A_y^{i,j}(0, z) = 0$. Since ψ is a solution of (2), $\partial_t \partial_{z^j} \psi^i$ exists and $\partial_t \partial_{z^j} \psi^i = \partial_{z^j} \partial_t \psi^i$. Therefore, $A_y^{i,j}$ is differentiable by t, and

$$\begin{aligned} \dot{A}_y^{i,j}(t, z) &= \partial_t \partial_{z^j} \psi^i(t; y, z) + \bar{g}_i^{m+j}(\Psi_y(t, z)) \\ &\quad + tD\bar{g}_i^{m+j}(\Psi_y(t, z))\dot{\Psi}_y(t, z) \\ &= \partial_{z^j} \dot{\psi}^i(t; y, z) + \bar{g}_i^{m+j}(\Psi_y(t, z)) \\ &\quad + t[\sum_{k=1}^{m} \partial_k \bar{g}_i^{m+j}(\Psi_y(t, z))\dot{\psi}^k(t; y, z) \\ &\quad + \sum_{\ell=1}^{n-m} \partial_{m+\ell} \bar{g}_i^{m+j}(\Psi_y(t, z)) z^\ell]. \end{aligned}$$

Further,

$$\begin{aligned} \partial_{z^j} \dot{\psi}^i(t; y, z) &= \partial_{z^j}[\sum_{\ell=1}^{n-m} -\bar{g}_i^{m+\ell}(x_1 + \psi(t; y, z), x_2 + tz) z^\ell] \\ &= -\bar{g}_i^{m+j}(\Psi_y(t, z)) \\ &\quad - \sum_{\ell=1}^{n-m} z^\ell \Big[\sum_{k=1}^{m} \partial_k \bar{g}_i^{m+\ell}(\Psi_y(t, z)) \partial_{z^j} \psi^k(t; y, z) \\ &\qquad + t\partial_{m+j} \bar{g}_i^{m+\ell}(\Psi_y(t, z))\Big], \end{aligned}$$

and thus,

$$\dot{A}_y^{i,j}(t,z) = t\sum_{\ell=1}^{n-m} z^\ell[\partial_{m+\ell}\bar{g}_i^{m+j}(\Psi_y(t,z)) - \partial_{m+j}\bar{g}_i^{m+\ell}(\Psi_y(t,z))]$$
$$+\sum_{k=1}^{m}\Big[t\partial_k\bar{g}_i^{m+j}(\Psi_y(t,z))\dot{\psi}^k(t;y,z)$$
$$-\sum_{\ell=1}^{n-m}\partial_k\bar{g}_i^{m+\ell}(\Psi_y(t,z))\partial_{z^j}\psi^k(t;y,z)z^\ell\Big].$$

Since

$$\dot{\psi}^k(t;y,z) = \sum_{\ell=1}^{n-m} -\bar{g}_k^{m+\ell}(\Psi_y(t,z))z_\ell$$

and

$$\partial_{z^j}\psi^k(t;y,z) = A_y^{k,j}(t,z) - t\bar{g}_k^{m+j}(\Psi_y(t,z)),$$

we have

$$\dot{A}_y^{i,j}(t,z) = -\sum_{k=1}^{m}\sum_{\ell=1}^{n-m} z^\ell\partial_k\bar{g}_i^{m+\ell}(\Psi_y(t,z))A_y^{k,j}(t,z)$$
$$+t\sum_{\ell=1}^{n-m} z^\ell\Big[\partial_{m+\ell}\bar{g}_i^{m+j}(\Psi_y(t,z)) - \partial_{m+j}\bar{g}_i^{m+\ell}(\Psi_y(t,z))$$
$$+\sum_{k=1}^{m}\{\partial_k\bar{g}_i^{m+\ell}(\Psi_y(t,z))\bar{g}_k^{m+j}(\Psi_y(t,z))$$
$$-\partial_k\bar{g}_i^{m+j}(\Psi_y(t,z))\bar{g}_k^{m+\ell}(\Psi_y(t,z))\}\Big]$$
$$= -\sum_{k=1}^{m}\sum_{\ell=1}^{n-m} z^\ell\partial_k\bar{g}_i^{m+\ell}(\Psi_y(t,z))A_y^{k,j}(t,z),$$

by condition (C-q) at x. Hence, $A_y^{i,j}(t,z)$ is a solution of the following initial value problem:

$$\begin{aligned}\dot{A}_y^{i,j}(t,z) &= -\textstyle\sum_{k=1}^{m}\sum_{\ell=1}^{n-m} z^\ell\partial_k\bar{g}_i^{m+\ell}(\Psi_y(t,z))A_y^{k,j}(t,z),\\ A_y^{i,j}(0,z) &= 0.\end{aligned} \quad (3)$$

Clearly, $A_y(t, z) \equiv 0$ is a solution of this problem. Therefore, by the uniqueness of the solution of the ODE, we have $A_y(t, z) \equiv 0$. Thus, we have proved that for any i,

$$\bar{g}_i(\Psi_y(t, z))^T D_z \Psi_y(t, z) = 0^T.$$

Now, suppose that $(v, w) \in x + U_5$, and let $(y, z) = (v - x_1, w - x_2)$. Then, $(y, z) \in U_5 \subset U_4$, and thus,

$$y - \psi(1, \hat{u}(y, z), z) = F(y, z, \hat{u}(y, z)) = 0,$$

which implies that $y = \psi(1, \hat{u}(y, z), z)$. Since $(y, z) \in U_5$, we have $(\hat{u}(y, z), z) \in U_6 \subset U_4$. Hence, $(v, w) = \Psi_{\hat{u}(y,z)}(1, z)$, and thus,

$$\bar{g}_i(v, w)^T D_z \Psi_{\hat{u}(y,z)}(1, z) = 0^T.$$

Next, suppose that $u \in V_6$ and choose any $z \in W_6$. Then, $(u, z) \in U_6$, and thus, there exists $(y, z) \in U_5$ such that $\hat{u}(y, z) = u$. Since $y = \psi(1, u, z)$, we have $\Psi_u(1, z) - x = (y, z)$. Hence, $\hat{u}(\Psi_u(1, z) - x) = u$. On differentiating both sides by z, we have

$$D\hat{u}(\Psi_u(1, z) - x) D_z \Psi_u(1, z) = 0,$$

for any $(u, z) \in U_6$.

Last, for any $(v, w) \in (x + U_5) \cap (x + U_6)$, define $(y, z) = (v - x_1, w - x_2)$ and $u(v, w) = \hat{u}(y, z)$. We have proved that $\bar{g}_i(v, w)^T D_z \Psi_{u(v,w)}(1, z) = 0^T$ for any i. Moreover, we have $z \in W_6$, $u(v, w) \in V_6$, and $\Psi_{u(v,w)}(1, z) - x = (y, z)$. Thus, $Du(v, w) D_z \Psi_{u(v,w)}(1, z) = 0^T$. Hence, both span$\{\bar{g}_1(v, w), ..., \bar{g}_m(v, w)\}$ and span$\{Du^1(v, w), ..., Du^m(v, w)\}$ are the orthogonal complement of the space span$\{D_{z^1} \Psi_{u(v,w)}(1, z), ..., D_{z^{n-m}} \Psi_{u(v,w)}(1, z)\}$. Therefore,

$$\text{span}\{\bar{g}_1(v, w), ..., \bar{g}_m(v, w)\} = \text{span}\{Du^1(v, w), ..., Du^m(v, w)\},$$

and thus,

$$\text{span}\{g_1(v, w), ..., g_m(v, w)\} = \text{span}\{Du^1(v, w), ..., Du^m(v, w)\},$$

which completes the proof of 2). ■

3.4. Proof of 3)

Since $\{g_1, ..., g_m\}$ satisfies condition (A-$(q + 1)$) at x, there exist open U such that $x \in U$ and $u : U \to \mathbb{R}^m$ such that u is C^{q+1}-class and span$\{g_1, ..., g_m\}$ = span$\{Du^1, ..., Du^m\}$. By Lemma 1, it suffices to show that $\{Du^1, ..., Du^m\}$ satisfies condition (B-q). In fact, $D^2 u^i$ is symmetric for any i, and thus, $h_{i,j} \equiv 0$ satisfies our desired condition. This completes the proof of 3). ■

References

1. Antonelli, G. B.: Sulla Teoria Matematica dell' Economia Politica. Tipografia del Folchetto, Pisa (1886)
2. Auslander, L., Mackenzie, R. E.: Introduction to Differential Manifolds. Dover Publications (2009)
3. Debreu, G.: Smooth preferences. Econometrica **40**, 603–615 (1972)
4. ————: Smooth preferences: a corrigendum. Econometrica **44**, 831–832 (1976)
5. Hicks, N. J.: Notes on Differential Geometry. Van Nostrand Reinhold, U.S. (1965)
6. Hurwicz, L.: On the problem of integrability of demand functions. In: Chipman, J. S., Hurwicz, L., Richter, M. K., Sonnenschein, H. (eds.) Preference, Utility and Demand, ch.9. Harcourt Brace Jovanovich, New York (1971)
7. Kosinski, A. A.: Differential Manifolds. Academic Press (1993)
8. Matsushima, Y.: Differentiable Manifolds. Marcel Dekker (1972)
9. Pontryagin, L. S.: Ordinary Differential Equation. Addison-Wesley (1962)
10. Samuelson, P. A.: The problem of integrability in utility theory. Economica **17**, 355–385 (1950)
11. Sternberg, S.: Lectures on Differential Geometry. American Mathematical Society (1999)

Adv. Math. Econ. 16, 53–84 (2012)

Advances in MATHEMATICAL ECONOMICS

A remark on credit risk models and copula

Shigeo Kusuoka[1,*] and Takenobu Nakashima[2]

[1] Graduate School of Mathematical Sciences, The University of Tokyo
(e-mail: kusuoka@ms.u-tokyo.ac.jp)

[2] Nomura Securities Company, Graduate School of Mathematical Sciences, The University of Tokyo
(e-mail: nakashima-0h1m@jp.nomura.com)

Received: June 2, 2011
Revised: September 13, 2011

JEL classification: G12

Mathematics Subject Classification (2010): 60G44, 91G40

Abstract. Copula models with finite parameters are widely used to describe the joint distribution of default times. But it is not clear whether these copula models are dynamically consistent. The authors show that the set of copula models that are dynamically consistent and satisfy some technical regularity conditions, is a set of the first category in the Baire sense in a certain space of copula functions with finite parameters.

Key words: copula, credit risk, default times

1. Introduction

Recently, copula models are widely used to describe the joint distribution of stopping times (such as borrower default times). But it is not clear whether these copula models are dynamically consistent. For consistency, there is an enormous family of corresponding martingale restrictions. Specifically, the

* Research supported by the 21st century COE project, Graduate School of Mathematical Sciences, The University of Tokyo.

process describing the conditional probability, as time passes and new information is gathered, of any specific joint default scenario must be a martingale, by the law of iterated expectations. The scenario could be, for example, that borrowers A and B default by January 1, 2015, while borrower C survives until at least June 1, 2015.

In a static setting, the copula has been conveniently used to parameterize the likelihoods of such scenarios, particularly for purposes of analyzing collateralized debt obligations and for capital-sufficiency analysis of credit portfolios. But the same copulas have been applied in dynamic settings, routinely, and nobody to this point has ever checked whether this can be done consistently in a dynamic setting. This is critical for dynamic risk management purposes and obviously for checking the validity of the chosen copula function.

Björk-Christernsen [3] considered the relationship between a family of forwardrate curves parameterized by finite factors and a dynamical interest rate model free of arbitrage, and showed that there are some analytic constraint conditions for a family of forwardrate curves which comes from a dynamical interest rate model free of arbitrage.

In the present paper, we study the relationship between a family of copula functions parameterized by finite dimensional parameters and dynamical default time models. Then we show that the set of copula models that are consistent in this dynamic sense@ and satisfy some technical regularity conditions, is extremely rare (a set of the first category in the Baire sense). In economic theory, this notion of rareness is described by saying that the consistency condition is "generically" not satisfied. The paper also analyzes consistency of the family of Gumbel copulas and concludes that it is not consistent.

The setup in this paper is the following.

Let $(\Omega, \mathcal{F}, P)$ be a complete probability space, $W(t) = (W^k(t))_{k=1,\dots,d}$, $t \geqq 0$, be a d-dimensinal standard Wiener process, and $\mathcal{G}_t = \sigma\{W(s),\ s \in [0,t]\} \vee \mathcal{N}$, where $\mathcal{N} = \{B \in \mathcal{F};\ P(B) = 0 \text{ or } 1\}$. Let $N \geqq 2$, $\tau_i : \Omega \to [0,\infty)$, $i = 1, \dots, N$, be random variables, and let $\mathcal{F}_t = \mathcal{G}_t \vee \sigma\{\tau_i \wedge t, i = 1, \dots, N\}$. Let $\xi_i : [0,\infty) \times \Omega \to [0,\infty)$, $i = 1, \dots, N$, be $\mathcal{G}$-progressively measurable processes.

First, we assume the following conditions.

$$\text{(SC)}\ \ (\prod_{i\in I} 1_{\{\tau_i > t\}}) P(\tau_i > t_i, i \in I | \mathcal{F}_t)$$
$$= (\prod_{i\in I} 1_{\{\tau_i > t\}}) E[\exp(-\sum_{i\in I} \int_t^{t_i} \xi_i(s)ds) | \mathcal{G}_t]\ a.s.$$

for any $I \subset \{1, \ldots, N\}$ and $t, t_i \in [0, \infty), i \in I$ with $t \leqq \min_{i \in I} t_i$.
(PO) For any $t \geqq 0$,

$$P(\bigcap_{i=1}^{N}\{\tau_i > t\})|\mathcal{G}_t] > 0 \; a.s.$$

We also assume the following technical assumptions.
(A-1) For any $T > 0$,

$$\sum_{i=1}^{N}\int_0^T E[\xi_i(t)^4]dt < \infty.$$

(A-2) For any $i = 1, \ldots, N$,

$$\int_0^\infty \xi_i(t) = \infty \quad a.s. \text{ and } \int_a^b \xi_i(t) > 0 \quad a.s. \text{ for any } a, b > 0 \text{ with } b > a.$$

(A-3) $\displaystyle\sum_{i=1}^{N}\int_0^\infty (1+t)^2 E[\xi_i(t)^2 \exp(-2\int_0^t \xi_i(s)ds)]dt < \infty.$

Let $\theta : [0, \infty) \times \Omega \to \mathbf{R}^M$ be a $\mathcal{G}$-Ito process, i.e., θ is $\mathcal{G}$-progressively measurable, $\theta(t, \omega)$ is continuous in t for all $\omega \in \Omega$, and there are $\mathbf{R}^M$-valued $\mathcal{G}$-progressively measurable processes η_k, $k = 1, \ldots, d$, and b satisfying

$$P(\sum_{k=1}^{d}\int_0^T |\eta_k(t)|^2 dt + \int_0^T |b(t)|dt < \infty) = 1, \text{ for any } T > 0,$$

and

$$\theta(t) = \theta(0) + \sum_{k=1}^{d}\int_0^t \eta_k(s)dW^k(s) + \int_0^t b(s)ds. \tag{1}$$

Let Θ be an open subset in $\mathbf{R}^M$ and $K \in C([0,1]^N \times \Theta; [0,1])$. We assume the following, moreover.
(A-4) $P(\theta(t) \in \bar{\Theta}$ for all $t \geqq 0) = 1$, where $\bar{\Theta}$ is the closure of Θ in $\mathbf{R}^M$.
(A-5) the support of probability law of $\theta(t, \omega)$ under $e^{-t}dt \otimes P(d\omega)$ contains a non-empty open set in Θ, i.e., there is a non-empty open set U_0 in Θ such that for any $\theta_0 \in U_0$ and $\varepsilon > 0$

$$\int_0^\infty P(|\theta(t) - \theta_0| < \varepsilon)\, e^{-t}dt > 0.$$

(CP) $K(\cdot,\theta) : [0,1]^N \to [0,1]$ is a copula function for any $\theta \in \Theta$, and

$$(\prod_{i=1}^{N} 1_{\{\tau_i>t\}})1_{\Theta}(\theta(t))P(\tau_i > t_i, i = 1,\dots,N|\mathcal{F}_t)$$

$$= (\prod_{i=1}^{N} 1_{\{\tau_i>t\}})1_{\Theta}(\theta(t))K(P(\tau_1 > t_1|\mathcal{F}_t),\dots,(P(\tau_N > t_N|\mathcal{F}_t),\theta(t))\ a.s.$$

for any $t, t_1, \dots, t_N > 0$ with $t < \min_{i=1,\dots,N} t_i$.

We call a family $((\Omega,\mathcal{F},P),(W_t^k)_{k=1,\dots,d},(\tau_i)_{i=1,\dots,N},(\xi_i(t))_{i=1,\dots,N},\theta(t),\Theta,K)$ satisfying the above assumptions a dynamical default time copula model, and we call K the associated family of copula functions to this model.

Definition 1. *Let Θ be an open subset in $\mathbf{R}^M$. We say that $K \in C([0,1]^N \times \Theta; [0,1])$ is an admissible family of copula functions, if there is a dynamical default time copula model and K is the associated family of copula functions to the model.*

The purpose of the present paper is to show that there are some analytic constraint conditions for an admissible family of copula functions. For example we will prove the following.

Let $N, M \geqq 1$, Θ be a non-void open subset in $\mathbf{R}^M$. Let $\mathcal{C}_{(N)}(\Theta)$ denote the subset of $C([0,1]^N \times \Theta; [0,1])$ consisting of elements K such that $K(\cdot,\theta) : [0,1]^N \to [0,1]$ is a copula function for any $\theta \in \Theta$, and $K|_{(0,1)^N\times\Theta}$ is a C^∞ function.

Let D_n be an increasing sequence of compact subsets in Θ such that $\bigcup_{n=1}^{\infty} D_n = \Theta$. Then we can regard $\mathcal{C}_{(N)}(\Theta)$ as a Polish space with a metric function dis given by

$$dis(K_1, K_2)$$

$$= \sum_{n=1}^{\infty} 2^{-n} \wedge \sup\{|K_1(x,\theta) - K_2(x,\theta)|;\ x \in [0,1]^N,\ \theta \in D_n\}$$

$$+ \sum_{n=1}^{\infty} 2^{-n} \wedge (\sum_{\alpha_1,\dots,\alpha_{N+M}=0}^{n} \sup\{| \frac{\partial^{\alpha_1+\cdots\alpha_{N+M}}(K_1 - K_2)}{\partial x_1^{\alpha_1}\dots\partial x_N^{\alpha_N}\partial\theta_1^{\alpha_{N+1}}\partial\dots\theta_M^{\alpha_{N+M}}}(x,\theta);$$

$$x \in [1/4n, 1 - 1/4n]^N,\ \theta \in D_n\}).$$

Our main result is the following.

Theorem 2. *Let $N \geqq 3$, $M \geqq 1$, and Θ be a non-void open subset in $\mathbf{R}^M$. Then the subset of $\mathcal{C}_{(N)}(\Theta)$ whose elements are admissible families of copula functions is a set of the first category in Baire's sense.*

We also show that a family of Gumbel copula functions of 3 variables is not admissible by relying on numerical computation in Sect. 7.

Acknowledgement
The authors thank the referee for useful suggestions and comments.

2. Preliminary results

Let $\xi : [0,\infty) \times \Omega \to [0,\infty)$ be a $\mathcal{G}$-progressively measurable satisfying the following three conditions.

(B-1) $\int_0^T E[\xi(t)^4]dt < \infty$ for any $T > 0$.

(B-2) $\int_0^\infty \xi(t)dt = \infty$ $a.s.$, and $\int_a^b \xi(t)dt > 0$ $a.s.$ for any $b > a \geqq 0$.

(B-3) $E[\int_0^\infty (1+t)^2 \xi(t)^2 \exp(-2\int_0^t \xi(r)dr)dt] < \infty$.

For each $s \geqq 0$, let $\{M(t,s); t \geqq 0\}$ is a continuous martingale given by

$$M(t,s) = E[\exp(-\int_0^s \xi(r)dr)|\mathcal{G}_t], \qquad t \geqq 0.$$

Proposition 3. *There is $f : [0,\infty) \times [0,\infty) \times \Omega \to (0,\infty)$ satisfying the following.*

(1) *For any $t, s \geqq 0$,*

$$f(t,s) = E[\exp(-\int_{t\wedge s}^s \xi(r)dr)|\mathcal{G}_t] = \exp(\int_0^{t\wedge s} \xi(r)dr)M(t,s) \qquad a.s.$$

(2) *For any $\omega \in \Omega$ $f(\cdot, * : \omega) : [0,\infty) \times [0,\infty) \to [0,\infty)$ is continuous.*

(3) *If $s_2 > s_1 > t$, then $f(t,s_1,\omega) > f(t,s_2,\omega) > 0$, $\omega \in \Omega$.*

(4) *For any $t \geqq 0$*

$$f(t,t) = 1, \quad \lim_{s\uparrow\infty} f(t,s) = 0.$$

Proof. Note that for $0 \leqq s_1 \leqq s_2$

$$E[\sup_{t\in[0,\infty)} |M(t,s_1) - M(t,s_2)|^4] \leqq 4E[|M(s_1,s_1) - M(s_2,s_2)|^4]$$

$$\leqq 4E[|\int_{s_1}^{s_2} \xi(r)dr|^4] \leqq 4(s_2 - s_1)^3(\int_{s_1}^{s_2} E[\xi(r)^4]dr).$$

So by Kolmogorov's continuity theorem and the assumption (B-1), we see that there is a $\tilde{M} : [0,\infty) \times [0,\infty) \times \Omega \to [0,\infty)$ such that $\tilde{M}(\cdot, *, \omega) \to [0,\infty)$ is continuous and $P(\tilde{M}(t,s) = M(t,s)) = 1$, $t, s \geqq 0$. Let

$$\tilde{f}(t,s) = \exp(\int_0^{t\wedge s} \xi(r)dr)\tilde{M}(t,s) \qquad t,s \geqq 0.$$

Then $\tilde{f}(t,s)$ is continuous in (t,s). Let $0 \leqq s_1 < s_2$. Then

$$\tilde{f}(t,s_1) - \tilde{f}(t,s_2) = \exp(\int_0^t \xi(r)dr)(\tilde{M}(t,s_1) - \tilde{M}(t,s_2)), \qquad t \in [0,s_1].$$

By the assumption (B-2), we have

$$\begin{aligned}&\tilde{M}(s_1,s_1) - \tilde{M}(s_1,s_2)\\ &= \exp(-\int_0^{s_1} \xi(r)dr)E[1 - \exp(-\int_{s_1}^{s_2} \xi(r)dr))|\mathcal{G}_{s_1}] > 0 \; a.s.\end{aligned}$$

Let $\tau = \inf\{t \geqq 0;\ \tilde{M}(t,s_1) - \tilde{M}(t,s_2) = 0\} \wedge s_1$. Then we see that

$$\tilde{M}(\tau,s_1) - \tilde{M}(\tau,s_2) = E[\tilde{M}(s_1,s_1) - \tilde{M}(s_1,s_2)|\mathcal{G}_\tau] > 0, \qquad a.s.$$

So we see that

$$\inf_{t\in[0,s_1]} (\tilde{M}(t,s_1) - \tilde{M}(t,s_2)) > 0 \; a.s.,$$

and so we have

$$\inf_{t\in[0,s_1]} (\tilde{f}(t,s_1) - \tilde{f}(t,s_2)) > 0 \; a.s.$$

for any $s_1, s_2 \in \mathbf{Q}$ with $s_2 > s_1 \geqq 0$. So there is an $\Omega_1 \in \mathcal{F}$ with $P(\Omega_1) = 1$ such that $\tilde{f}(t,s_1,\omega) > \tilde{f}(t,s_2,\omega)$ for any $\omega \in \Omega_1$, $s_1, s_2 \in \mathbf{Q}$ with $s_2 > s_1 > 0$ and $t \in [0,s_1)$. Since $\tilde{f}(t,s)$ is continuous in (t,s), we see that $\tilde{f}(t,s)$ is non-increasing in s. So we see that $\tilde{f}(t,s_1,\omega) > \tilde{f}(t,s_2,\omega)$ for any $\omega \in \Omega_1$, $t, s_1, s_2 \in [0,\infty)$ with $s_2 > s_1 > t$. Similarly we can show that there is an $\Omega_2 \in \mathcal{F}$ with $P(\Omega_2) = 1$ such that $\tilde{f}(t,s,\omega) > 0$ for any $\omega \in \Omega_2$, $t, s \in [0,\infty)$.

We see that

$$E[\varliminf_{s\to\infty} \tilde{f}(t,s)] \leqq \varliminf_{s\to\infty} E[\tilde{f}(t,s)] = \varliminf_{s\to\infty} E[\exp(-\int_t^s \xi(r)dr] = 0.$$

Since $\lim_{s\to\infty} \tilde{f}(t,s,\omega)$ exists for $\omega \in \Omega_1$, we see that $\lim_{s\to\infty} \tilde{f}(t,s,\omega) = 0$ a.s. Also, it is easy to see $\tilde{f}(t,t) = 1$ $a.s.$ Therefore we can take a good version f of $\tilde{f}$ satisfying the assertion. ∎

Proposition 4. *There exist* $\hat{\sigma}_k : [0,\infty] \times [0,\infty) \times \Omega \to \mathbf{R}$, $k = 1,\ldots,d$, *satisfying the following.*

(1) $\hat{\sigma}_k(t,\cdot,\omega) : [0,\infty] \to \mathbf{R}$, $k = 1,\ldots,d$, *is continuous for any* $t \in [0,\infty)$ *and* $\omega \in \Omega$. *Moreover,* $\hat{\sigma}_k(t,s,\omega) = 0$, $t \geqq s$, *and* $\lim_{s\to\infty} \hat{\sigma}_k(t,s,\omega) = 0$ *for any* $t \in [0,\infty)$ *and* $\omega \in \Omega$.

(2) $\hat{\sigma}_k(\cdot,s) : [0,\infty) \times \Omega \to \mathbf{R}$, $k = 1,\ldots,N$, *is* $\mathcal{G}$*-progressively measurable for any* $s \geqq 0$ *and*

$$M(t,s) = M(0,s) + \sum_{k=1}^{d} \int_0^t \hat{\sigma}_k(r,s)dW^k(r), \quad t \geqq 0, \ a.s.$$

for any $s > 0$.

Proof. For each $s \geqq 0$, let $N(t,s), t \in [0,\infty)$ be a continuous martingale given by

$$N(t,s) = E[\xi(s)\exp(-\int_0^s \xi(r)dr)|\mathcal{G}_t].$$

By Ito's representation theorem, we see that for any $s \geqq 0$ there exist $\mathcal{G}$-progressively measurable processes $c_k(\cdot,s) : [0,\infty) \times \Omega \to \mathbf{R}$, $k = 1,\ldots,d$, such that

$$N(t,s) = N(0,s) + \sum_{k=1}^{d} \int_0^t c_k(r,s)dW^k(r), \quad t \geqq 0.$$

Since the map from $[0,\infty)$ to $L^2(\Omega,\mathcal{F},P)$ coresponding s to $N(t,s)$ is measurable, we may assume that $c_k : [0,\infty) \times [0,\infty) \times \Omega \to \mathbf{R}$ is measurable. Note that

$$N(0,s)^2 + \sum_{k=1}^{d} \int_0^\infty E[c_k(r,s)^2]dr$$

$$= \lim_{t\to\infty} E[N(t,s)^2] \leqq E[\xi(s)^2 \exp(-2\int_0^s \xi(r)dr)].$$

Therefore by the assumption (B-3), we see that

$$E[\int_{[0,\infty)\times[0,\infty)} (1+s)^2 c_k(r,s)^2 drds] < \infty, \qquad k = 1,\ldots,d.$$

Let us define $\tilde{\sigma}_k : [0,\infty) \times [0,\infty) \times \Omega \to \mathbf{R}$, $k = 1,\ldots,N$, by

$$\tilde{\sigma}_k(t,s) = \begin{cases} -\int_0^s c_k(t,u)du, & \text{if } \int_{[0,\infty)}(1+u)^2 c_k(t,u)^2 ds < \infty, \\ 0, & \text{otherwise.} \end{cases}$$

Then we see that $\tilde{\sigma}_k(\cdot, s) : [0, \infty) \times \Omega \to \mathbf{R}$, is $\mathcal{G}$-progressively measurable for any $s \geqq 0$ and $\tilde{\sigma}_k(t, \cdot) : [0, \infty) \to \mathbf{R}$ is continuous. Also, by stochastic Fubini's theorem, we have

$$-\int_0^s N(0,u)du + \sum_{k=1}^d \int_0^t \tilde{\sigma}_k(r,s)dW^k(r)$$

$$= -\int_0^s N(0,u)du - \sum_{k=1}^d \int_0^t (\int_0^s c_k(r,u)du)dW^k(r)$$

$$= -\int_0^s N(t,u)du = E[\exp(-\int_0^s \xi(u)du) - 1|\mathcal{G}_t] = M(t,s) - 1 \qquad a.s.$$

So we see that

$$M(t,s) = M(0,s) + \sum_{k=1}^d \int_0^t \tilde{\sigma}_k(r,s)dW^k(r).$$

Note that for $0 < s_1 < s_2 < \infty$, we have

$$|\tilde{\sigma}_k(t,s_1) - \tilde{\sigma}_k(t,s_2)|^2 \leqq (\int_{s_1}^{s_2} |c_k(t,u)|du)^2$$

$$\leqq (\int_{s_1}^{s_2} (1+u)^{-2}du)(\int_{s_1}^{s_2} (1+u)^2 c_k(t,u)^2 du) \leqq \int_{s_1}^{\infty} (1+u)^2 c_k(t,u)^2 du.$$

So we have

$$E[\int_{[0,\infty)} dt (\sup_{s_1,s_2>s} |\tilde{\sigma}_k(t,s_1) - \tilde{\sigma}_k(t,s_2)|^2)]$$

$$\leqq E[\int_{[0,\infty)\times[s,\infty)} (1+u)^2 c_k(r,u)^2 drdu] \to 0, \qquad s \to \infty.$$

Therefore we see that

$$\sup_{s_1,s_2>s} |\tilde{\sigma}_k(t,s_1) - \tilde{\sigma}_k(t,s_2)|^2 \to 0, \qquad s \to \infty \qquad dt \otimes P(d\omega) - a.e.(t,\omega).$$

This implies that $\tilde{\sigma}_k(t,s)$ converges as $s \to \infty$ for $dt \otimes P(d\omega) - a.e.(t,\omega)$.

Also, we see by (B-2) that

$$E[\int_0^\infty (\varliminf_{s\to\infty} \tilde{\sigma}_k(t,s)^2)dt] \leqq \varliminf_{s\to\infty} E[\int_0^\infty \tilde{\sigma}_k(t,s)^2 dt]$$

$$\leqq \varliminf_{s\to\infty} E[\exp(-2\int_0^s \xi_i(u)du)] = 0.$$

Thus we see that $\tilde{\sigma}_k(t,s) \to 0,\ s \to \infty$ for $dt \otimes P(d\omega) - a.e.(t,\omega)$.

Let $\hat{\sigma}_k$, $k = 1, \dots, d$, be given by

$$\hat{\sigma}_k(t,s) = \begin{cases} \tilde{\sigma}_k(t,s), & \text{if } \tilde{\sigma}_k(t,s) \to 0, \text{ as } s \to \infty, \\ 0, & \text{otherwise,} \end{cases}$$

Then we have our assertion. ■

By Ito's formula, we have

$$f(t,s) = f(0,s) + \int_0^{t\wedge s} \xi(r) f(r,s)dr$$
$$+ \sum_{k=1}^d \int_0^t \exp\left(\int_0^{r\wedge s} \xi(u)du\right)\hat{\sigma}_k(r,s)dW^k(r), \quad t \geqq 0,$$

for any $s \geqq 0$. So we have the following as a corollary to Proposition 4.

Corollary 5. *There exist* $\tilde{\sigma}_k : [0,\infty] \times [0,\infty) \times \Omega \to \mathbf{R}$, $k = 1, \dots, d$, *such that*

(1) $\tilde{\sigma}_k(t,\cdot,\omega) : [0,\infty] \to \mathbf{R}$, $k = 1, \dots, d$, *is continuous for any* $t \in [0,\infty)$ *and* $\omega \in \Omega$. *Moreover,* $\tilde{\sigma}_k(t,s,\omega) = 0$, $t \geqq s$, *and* $\lim_{s\to\infty} \tilde{\sigma}_k(t,s,\omega) = 0$ *for any* $t \in [0,\infty)$ *and* $\omega \in \Omega$.

(2) $\tilde{\sigma}_{i,k}(\cdot,s) : [0,\infty) \times \Omega \to \mathbf{R}$, $k = 1, \dots, N$, *is* $\mathcal{G}$*-progressively measurable for any* $s \geqq 0$ *and*

$$f(t,s) = f(0,s) + \int_0^{t\wedge s} \xi(r) f(r,s)dr$$
$$+ \sum_{k=1}^d \int_0^t \tilde{\sigma}_k(r,s)dW^k(r), \quad t \geqq 0, \ a.s.$$

for any $s > 0$.

By Proposition 3, we have the following immediately.

Proposition 6. *Let* $T : [0,\infty) \times (0,1] \times \Omega \to [0,\infty]$ *be given by*

$$T(t,x) = \inf\{s \geqq t;\ f(t,s) < x\}, \qquad x \in (0,1].$$

Then $T(t,\cdot,\omega) : (0,1] \to [0,\infty)$ *is continuous and strictly decreasing and* $\lim_{x\downarrow 0} T(t,x,\omega) = \infty$ *for any* $t \geqq 0$ *and* $\omega \in \Omega$.

Now let $X : [0,\infty) \times [0,\infty) \times (0,1] \times \Omega \to (0,1]$ be given by

$$X(t,s,x) = f(t \vee s, T(s,x)), \qquad t,s \geqq 0,\ x \in (0,1].$$

Then we see that $\lim_{x\to 0} X(t,s,x,\omega) = 0$. So by defining $X(t,s,0) = 0$, we can define $X : [0,\infty) \times [0,\infty) \times [0,1] \times \Omega \to [0,1]$ such that $X(\cdot, *, **, \omega) : [0,\infty) \times [0,\infty) \times [0,1] \to [0,1]$ is continuous for any $\omega \in \Omega$, $X(t,s,\cdot,\omega) : [0,1] \to [0,1]$ is continuous and non-decreasing, and

$$f(t \vee s, r) = X(t, s, f(s,r)), \qquad r \geqq s \geqq 0,\ t \geqq 0.$$

Then we see that $X(t,t,x) = x$, and for $t \geqq s \geqq r \geqq 0$,

$$\begin{aligned} X(t,s,X(s,r,x)) &= f(t, T(s, f(s, T(r,x)))) = f(t, T(r,x) \vee s) \\ &= f(t, T(r,x)) = X(t,r,x). \end{aligned}$$

Let $Y(t,s) = \inf\{x \in [0,1];\ X(t,s,x) = 1\}$. Then we see that $T(s,x) \geqq t$ iff $x \geqq Y(t,s)$, and that $Y(t,s)$ is $\mathcal{G}_s$-measurable.

3. A remark on support

Let $(\Omega, \mathcal{F}, P)$ be a probability measure, Θ be a non-empty open set in $\mathbf{R}^M$, and $\mathcal{M}_0$ be a Polish space. Also, let $\xi : [0,\infty) \times \Omega \to [0,\infty)$, $\theta : [0,\infty) \times \Omega \to \bar{\Theta}$, and $Y : [0,\infty) \times \Omega \to \mathcal{M}_0$ be measurable processes. Remind that $\bar{\Theta}$ is the closure of Θ in $\mathbf{R}^M$. We assume that $\theta(\cdot,\omega) \to \bar{\Theta}$ is continuous for all $\omega \in \Omega$ and that $P(\int_a^b \xi(t)dt > 0) = 1$ for any $a, b \geqq 0$ with $a < b$.

Let $\tilde{\Omega} = [0,\infty) \times \Omega$. Let ν_0 be a probability measure on $[0,\infty)$ given by $\nu_0(dt) = e^{-t}dt$, and ν be a probability measure on $(\tilde{\Omega}, \mathcal{B}([0,\infty)) \times \mathcal{F})$ given by $\nu = \nu_0 \otimes P$. Then ξ, (resp. θ, Y) can be regarded as a $[0,\infty)$ (resp. Θ, $\mathcal{M}_0$)-valued random variable defined in a probability space $(\tilde{\Omega}, \mathcal{B}([0,\infty)) \times \mathcal{F}, \nu)$.

Let μ be a probability law of (ξ, Y, θ) and μ_θ be a probability law of θ unde ν. Then μ and μ_θ be probability measures on $[0,\infty) \times \mathcal{M}_0 \times \bar{\Theta}$ and $\bar{\Theta}$ respectively. Let Γ and Γ_θ be the support of probability measures μ and μ_θ respectively. Then Γ and Γ_θ are closed subsets of $[0,\infty) \times \mathcal{M}_0 \times \bar{\Theta}$ and $\bar{\Theta}$ respectively. Let $\pi : [0,\infty) \times \mathcal{M}_0 \times \bar{\Theta} \to \bar{\Theta}$ be a natural projection and let $\Gamma_0 = \pi((0,\infty) \times \mathcal{M}_0 \times \Theta)$.

Then we have the following.

Proposition 7. *The closure of* Γ_0 *contains* $\Gamma_\theta \cap \Theta$.

Proof. Let $\Phi : \tilde{\Omega} \to [0,\infty) \times \mathcal{M}_0 \times \bar{\Theta}$ be given by $\Phi(t,\omega) = (\xi(t,\omega), Y(t,\omega), \theta(t,\omega))$. Let $A = \Phi^{-1}(\Gamma)$. Then we have

$$1 = \nu(A) = \int_\Omega \nu_0(A_\omega) P(d\omega),$$

where $A_\omega = \{t \in [0,\infty);\ (t,\omega) \in A\}$. Let

$$B = \{\omega \in \Omega;\ \nu_0(A_\omega) = 1,\ \int_r^{r'} \xi(t,\omega)dt > 0 \text{ for any } r.r' \in \mathbf{Q} \text{ with } r < r' \}.$$

Then we see that $P(B) = 1$. Let $A' = A \cap ([0,\infty) \times B)$. Then we see that $\nu(A') = 1$. Let $\theta_0 \in \Gamma_\theta \cap \Theta$. Then for any $n \geqq 1$,

$$\nu(\{(t,\omega) \in A';\ |\theta(t,\omega) - \theta_0| < \frac{1}{2n}\}) > 0.$$

Therefore there is a $(t_n,\omega_n) \in A'$ such that $|\theta(t_n,\omega_n) - \theta_0| < 1/2n$. For any $m \geqq 1$, we see that $\int_{t_n}^{t_n+1/m} \xi(t,\omega_n)dt > 0$, and so there is a $s_{n,m} \in (t_n, t_n + 1/m) \cap A_{\omega_n}$ such that $\xi(s_{n,m},\omega_n) > 0$. Since $\theta(t,\omega_n)$ is continuous in t, we see that there is a $m(n) \geqq 1$ such that $|\theta(s_{n,m(n)},\omega_n) - \theta(t_n,\omega_n)| < 1/2n$. Now let $\xi_n = \xi(s_{n,m(n)},\omega_n)$, $\theta_n = \theta(s_{n,m(n)},\omega_n)$, and $y_n = Y(s_{n,m(n)},\omega_n)$. Then we see that $(\xi_n, y_n, \theta_n) \in \Gamma$, $\xi_n > 0$, and $|\theta_n - \theta_0| < 1/n$. Since Θ is open, $\theta_n \in \Theta$ for sufficiently large n. So we have our assertion. ■

4. Fundamental Relations

Let $(\Omega, \mathcal{F}, P, (W_t^k)_{k=1,\ldots,d}, (\tau_i)_{i=1,\ldots,N}, (\xi_i(t))_{i=1,\ldots,N}, \theta(t), \Theta, K)$ be a dynamical default time copula model as in Introduction. We also assume that $K|_{(0,1)^N\times\Theta}$ is C^2. We think about conditions which K must satisfy.

By Proposition 3, we see that there are $f_i : [0,\infty) \times [0,\infty) \times \Omega \to (0,\infty)$, $i = 1,\ldots,N$, such that

$$f_i(t,s) = E[\exp(-\int_{t\wedge s}^{s} \xi_i(r)dr)|\mathcal{G}_t] \quad a.s. \qquad t,s \geqq 0,$$

$f_i(\cdot,*,\omega) : [0,\infty) \times [0,\omega)$, are continuous for any $\omega \in \Omega$, $f_i(t,s_1,\omega) > f_i(t,s_2,\omega) > 0$ for $s_2 > s_1 > t$, $\omega \in \Omega$, and

$$f(t,t,\omega) = 1, \quad \lim_{s\uparrow\infty} f(t,s,\omega) = 0, \qquad t \geqq 0,\ \omega \in \Omega.$$

Also by Corollary 5, we see that there are $\tilde{\sigma}_{i,k} : [0,\infty]\times[0,\infty)\times\Omega \to \mathbf{R}$, $k = 1,\ldots,d$, $i = 1,\ldots,N$, satisfying the following.

(1) $\tilde{\sigma}_{i,k}(t,\cdot,\omega) : [0,\infty] \to \mathbf{R}$, $k = 1,\ldots,d$, is continuous for any $t \in [0,\infty)$ and $\omega \in \Omega$.
(2) $\tilde{\sigma}_{i,k}(t,s,\omega) = 0$, $t \geqq s$, and $\lim_{s\to\infty} \tilde{\sigma}_{i,k}(t,s,\omega) = 0$ for any $t \in [0,\infty)$ and $\omega \in \Omega$.
(3) $\sigma_{i,k}(\cdot,s) : [0,\infty) \times \Omega \to \mathbf{R}$, $k = 1,\ldots,N$, is $\mathcal{G}$-progressively measurable for any $s \geqq 0$.
(4) For any $s > 0$

$$f_i(t,s) = f_i(0,s) + \int_0^{t\wedge s} \xi_i(r) f_i(r,s)dr + \sum_{k=1}^d \int_0^t \tilde{\sigma}_{i,k}(r,s)dW^k(r).$$

Let $T_i : [0,\infty) \times (0,1) \times \Omega \to (0,\infty)$, $i = 1,\dots,N$, be given by

$$T_i(t,x) = \inf\{s \geqq t,\ f_i(t,s) \leqq x\}, \qquad x \in [0,1).$$

Then by Proposition 6 we see that $T_i(t,\cdot,\omega) : (0,1) \to (0,\infty)$ is continuous and strictly decreasing, $\lim_{x\downarrow 0} T_i(t,x,\omega) = \infty$ and $\lim_{x\uparrow 1} T_i(t,x,\omega) = 0$ for any $t \geqq 0$ and $\omega \in \Omega$. Let $\sigma_{i,k} : [0,\infty) \times (0,1) \times \Omega \to \mathbf{R}$, $i = 1,\dots,N$, $k = 1,\dots,d$, be given by

$$\sigma_{i,k}(t,x) = \tilde{\sigma}_{i,k}(t, T_i(t,x)) \qquad t \geqq 0,\ x \in (0,1).$$

Then we see that

$$\lim_{x\downarrow 0} \sigma_{i,k}(t,x) = 0, \qquad \lim_{x\uparrow 1} \sigma_{i,k}(t,x) = 0.$$

So we can extend this $\sigma_{i,k}$ as a function $\sigma_{i,k} : [0,\infty) \times [0,1] \times \Omega \to \mathbf{R}$, for which $\sigma_{i,k}(t,\cdot,\omega) : [0,1] \to \mathbf{R}$ is continuous for any $t \geqq 0$, $\omega \in \Omega$, and $\sigma_{i,k}(t,0) = \sigma_{i,k}(t,1) = 0$.

Let ν be a probability measure on $(0,\infty) \times \Omega$ given by $\nu(dt,d\omega) = e^{-t}dt \otimes P(d\omega)$. By the assumption (SC), we see that

$$1_{\{\tau_i > t\}} f_i(t,s) = 1_{\{\tau_i > t\}} P(\tau_i > s|\mathcal{F}_t) \quad a.s. \quad s \geqq t,\ i = 1,\dots,N,$$

and

$$\prod_{i=1}^N 1_{\{\tau_i > t\}} \exp(-\sum_{i=1}^N \int_0^t \xi_i(r)dr) P(\tau_1 > s_1,\dots,\tau_N > s_n|\mathcal{F}_t)$$

$$= \prod_{i=1}^N 1_{\{\tau_i > t\}} E[\exp(-\sum_{i=1}^N \int_0^{s_i} \xi_i(r)dr)|\mathcal{G}_t]\ a.s.$$

for $t \in [0, \min_{i=1,\dots,N} s_i]$. So by the assumption (CP) we have

$$\prod_{i=1}^N 1_{\{\tau_i > t\}} 1_\Theta(\theta(t)) \exp(-\sum_{i=1}^N \int_0^t \xi_i(r)dr) K(f_1(t,s_1),\dots,f_N(t,s_N),\theta(t))$$

$$= \prod_{i=1}^N 1_{\{\tau_i > t\}} 1_\Theta(\theta(t)) E[\exp(-\sum_{i=1}^N \int_0^{s_i} \xi_i(r)dr)|\mathcal{G}_t]\ a.s.$$

for $t \in [0, \min_{i=1,\dots,N} s_i]$. Therefore by the assumption (PO), we have

$$1_{\Theta}(\theta(t)) \exp(-\sum_{i=1}^{N} \int_0^t \xi_i(r)dr)K(f_1(t,s_1),\dots,f_N(t,s_N),\theta(t))$$
$$= 1_{\Theta}(\theta(t))E[\exp(-\sum_{i=1}^{N}\int_0^{s_i} \xi_i(r)dr)|\mathcal{G}_t]\ a.s.$$

for $t \in [0, \min_{i=1,\dots,N} s_i]$.

Now let us take a non-empty open set U in $\mathbf{R}^M$ such that $\bar{U} \subset \Theta$ and fix it for a while. For $T \geqq 0$, let $\tau_T^U : \Omega \to [0,\infty)$ be given by

$$\tau_T^U = \inf\{t \geqq T;\ \theta(t) \notin \bar{U}\} \wedge (T+1).$$

Then by the assumption (CP), we see that for any $s_1,\dots,s_N \geqq T$

$$1_U(\theta(T))\exp(-\sum_{i=1}^{N}\int_0^{t\wedge\tilde{\tau}} \xi_i(r)dr)K(f_1(t\wedge\tilde{\tau},s_1),\dots,f_N(t\wedge\tilde{\tau},s_N),$$
$$\theta(t\wedge\tilde{\tau})),\ t\in[T,T+1],$$

is a $\{\mathcal{G}_t\}_{t\in[T,T+1]}$-maringale, where $\tilde{\tau} = \tau_T^U \wedge \min_{i=1,\dots,N} s_i$.

Note that $\theta(t)$ is an Ito process satisfying Equation (1). Therefore, applying Ito's formula and comparing finite total variation process, we have for any $s_1,\dots,s_N \geqq T$

$$1_U(\theta(T))1_{[t,T+1]}(\tau_T^U \wedge \min_{i=1,\dots,N} s_i)\{-(\sum_{i=1}^{N}\xi_i(t))K(f_1(t,s_1),\dots,f_N(t,s_N),\theta(t))$$
$$+\sum_{i=1}^{N}\xi_i(t)f_i(t,s_i)\frac{\partial K}{\partial x_i}(f_1(t,s_1),\dots,f_N(t,s_N),\theta(t))$$
$$+\sum_{j=1}^{M} b^j(t)\frac{\partial K}{\partial \theta^j}(f_1(t,s_1),\dots,f_N(t,s_N),\theta(t))$$
$$+\frac{1}{2}\sum_{i,i'=1}^{N}\sum_{k=1}^{d}\tilde{\sigma}_{i,k}(t,s_i)\tilde{\sigma}_{i',k}(t,s_{i'})\frac{\partial^2 K}{\partial x_i \partial x_{i'}}(f_1(t,s_1),\dots,f_N(t,s_N),\theta(t))$$
$$+\frac{1}{2}\sum_{j,j'=1}^{M}\sum_{k=1}^{d}\eta_k^j(t)\eta_k^{j'}(t)\frac{\partial^2 K}{\partial \theta_j \theta_{j'}}(f_1(t,s_1),\dots,f_N(t,s_N),\theta(t))$$

$$+\sum_{i=1}^{N}\sum_{j=1}^{M}\sum_{k=1}^{d}\tilde{\sigma}_{i,k}(t,s_i)\eta_k^j(t)\frac{\partial^2 K}{\partial x_i\theta_j}(f_1(t,s_1),\ldots,f_N(t,s_N),\theta(t))\}=0 \tag{2}$$

for $\nu-a.e.(t,\omega)\in(T,T+1)\times\Omega$.

Note that the left hand side of Equation (2) is right continuous in $s_1,\ldots,s_N$. So we see that there is an $B_T^U\in\mathcal{B}((T,T+1))\times\mathcal{F}$ such that $\nu(((T,T+1)\times\Omega)\setminus B_T^U)=0$ and Equation (2) holds for all $(t,\omega)\in B_T^U$ and $s_1,\ldots,s_N\in[T,\infty)$.

Also, substituting $s_i=T_i(t,x_i)$, $i=1,\ldots,N$, to Equation (2), we see that for all $(t,\omega)\in B_T^U$

$$1_U(\theta(T))1_{[t,T+1]}(\tau_T^U)\{-(\sum_{i=1}^{N}\xi_i(t))K(x_1,\ldots,x_N,\theta(t))$$

$$+\sum_{i=1}^{N}\xi_i(t)x_i\frac{\partial K}{\partial x_i}(x_1,\ldots,x_N,\theta(t))+\sum_{j=1}^{M}b^j(t)\frac{\partial K}{\partial\theta^j}(x_1,\ldots,x_N,\theta(t))$$

$$+\frac{1}{2}\sum_{i,i'=1}^{N}\sum_{k=1}^{d}\sigma_{i,k}(t,x_i)\sigma_{i',k}(t,x_{i'})\frac{\partial^2 K}{\partial x_i\partial x_{i'}}(x_1,\ldots,x_N,\theta(t))$$

$$+\frac{1}{2}\sum_{j,j'=1}^{M}\sum_{k=1}^{d}\eta_k^j(t)\eta_k^{j'}(t)\frac{\partial^2 K}{\partial\theta_j\theta_{j'}}(x_1,\ldots,x_N,\theta(t))$$

$$+\sum_{i=1}^{N}\sum_{j=1}^{M}\sum_{k=1}^{d}\sigma_{i,k}(t,x_i)\eta_k^j(t)\frac{\partial^2 K}{\partial x_i\theta_j}(x_1,\ldots,x_N,\theta(t))\}=0 \tag{3}$$

for any $x_1,\ldots,x_N\in(0,1)$.

Let $J_2=\{(j,j')\in\{0,1,\ldots,M\}\times\{1,\ldots,M\};\ j\leqq j'\}$. We define linear operators $S_{ii'}^{(2)}$, $i,i'=1,\ldots,N$, $i<i'$, $S_{ij}^{(1)}$, $i=1,\ldots,N$, $j=0,1,\ldots,M$, and $S_{jj'}^{(0)}$, $(j,j')\in J_2$, from $C^2((0,1)^N\times\Theta)$ to $C((0,1)^N\times\Theta)$ by

$$(S_{ii'}^{(2)}F)(x,\theta)=\frac{\partial^2 F}{\partial x_i\partial x_{i'}}(x,\theta),\qquad 1\leqq i<i'\leqq N,$$

$$(S_{i0}^{(1)}F)(x,\theta)=\frac{\partial^2 F}{(\partial x_i)^2}(x,\theta),\qquad i=1,\ldots,N,$$

$$(S_{ij}^{(1)}F)(x,\theta)=\frac{\partial^2 F}{\partial x_i\partial\theta_j}(x,\theta),\qquad i=1,\ldots,N,\ j=1,\ldots,M,$$

$$(S^{(0)}_{jj'}F)(x,\theta) = \frac{\partial^2 F}{\partial\theta_j \partial\theta_{j'}}(x,\theta), \qquad 1 \leqq j \leqq j' \leqq N,$$

$$(S^{(0)}_{0j'}F)(x,\theta) = \frac{\partial F}{\partial\theta_{j'}}(x,\theta), \qquad 1 \leqq j' \leqq N,$$

for any $F \in C^2((0,1)^N \times \Theta)$.

Also, let us define $a^{(2)}_{ii'} : [0,\infty) \times [0,1] \times [0,1] \times \Omega \to \mathbf{R}$, $i, i' = 1, \ldots, N$, $i < i'$, $a^{(1)}_{ij} : [0,\infty) \times [0,1] \times \Omega \to \mathbf{R}$, $i = 1, \ldots, N$, $j = 0, 1, \ldots, M$, and $a^{(0)}_{jj'} : [0,\infty) \times [0,1] \times \Omega \to \mathbf{R}$, $(j, j') \in J_2$, by the following.

$$a^{(2)}_{ii'}(t, x_i, x_{i'}) = \sum_{k=1}^{d} \sigma_{i,k}(t, x_i)\sigma_{i',k}(t, x_{i'}), \qquad 1 \leqq i < i' \leqq N,$$

$$a^{(1)}_{i0}(t, x_i) = \frac{1}{2}\sum_{k=1}^{d} \sigma_{i,k}(t, x_i)^2 \qquad i = 1, \ldots, N,$$

$$a^{(1)}_{ij}(t, x_i) = \sum_{k=1}^{d} \hat{\sigma}_{i,k}(t, x_i)\eta_{j,k}(t), \qquad i = 1, \ldots, N, \ j = 1, \ldots, M,$$

$$a^{(0)}_{jj'}(t) = \sum_{k=1}^{d} \eta_{j,k}(t)\eta_{j',k}(t), \qquad j, j' = 1, \ldots, M, \ \text{with } j < j',$$

$$a^{(0)}_{jj}(t) = \frac{1}{2}\sum_{k=1}^{d} \eta_{j,k}(t)^2, \qquad j = 1, \ldots, M,$$

and

$$a^{(0)}_{0j'}(t) = b_{j'}(t), \qquad j' = 1, \ldots, M.$$

Then we have for all $(t,\omega) \in B^U_T$

$$1_U(\theta(T))1_{[t,T+1]}(\tau^U_T)\{\sum_{i=1}^{N} \xi_i(t)(x_i \frac{\partial K}{\partial x_i}(x,\theta(t)) - K(x,\theta(t)))$$

$$+ \sum_{1 \leqq i < i' \leqq N} a^{(2)}_{ii'}(t, x_i, x_{i'})(S^{(2)}_{ii'}K)(x,\theta(t)) + \sum_{i=1}^{N}\sum_{j=0}^{d} a^{(1)}_{ij}(t, x_i)(S^{(1)}_{ij}K)(x,\theta(t))$$

$$+ \sum_{(j,j') \in J_2} a^{(0)}_{jj'}(t)(S^{(0)}_{jj'}K)(x,\theta(t))\} = 0, \qquad x_1, \ldots, x_N \in (0,1). \qquad (4)$$

Now let U_n, $n = 1, 2, \ldots$, be non-empty open sets in $\mathbf{R}^M$ such that the closure of U_n is contained in Θ for each n, and $\bigcup_{n=1}^{\infty} U_n = \Theta$. Since $\theta(t)$ is continuous in t, we see that

$$\{(t, \omega) \in (0, \infty) \times \Omega;\ \theta(t, \omega) \in \Theta\}$$

$$= \bigcup_{T \in \mathbf{Q}_{\geqq 0}} \bigcup_{n=1}^{\infty} \{(t, \omega) \in (T, T+1) \times \Omega;\ \theta(T, \omega) \in U_n,\ t \leqq \tau_T^{U_n}(\omega)\}.$$

So let

$$B_0 = \bigcup_{T \in \mathbf{Q}_{\geqq 0}} \bigcup_{n=1}^{\infty} (B_T^{U_n} \cap \{(t, \omega) \in (T, T+1) \times \Omega;\ \theta(T, \omega) \in U_n,\ t \leqq \tau_T^{U_n}(\omega)\})$$

and $B_1 = B_0 \cup \{(t, \omega) \in (0, \infty) \times \Omega;\ \theta(t, \omega) \notin \Theta\}$. Then we see that $\nu(B_1) = 1$. Also, we see that for all $(t, \omega) \in B_1$

$$1_\Theta(\theta(t))\{\sum_{i=1}^N \xi_i(t)(x_i \frac{\partial K}{\partial x_i}(x, \theta(t)) - K(x, \theta(t)))$$

$$+ \sum_{1 \leqq i < i' \leqq N} a_{ii'}^{(2)}(t, x_i, x_{i'})(S_{ii'}^{(2)} K)(x, \theta(t))$$

$$+ \sum_{i=1}^N \sum_{j=0}^d a_{ij}^{(1)}(t, x_i)(S_{ij}^{(1)} K)(x, \theta(t))$$

$$+ \sum_{(j, j') \in J_2} a_{jj'}^{(0)}(t)(S_{jj'}^{(0)} K)(x, \theta(t))\} = 0, \qquad x_1, \ldots, x_N \in (0, 1). \tag{5}$$

Let $\mathcal{C}_2$ be the set of continuous functions $a : [0, 1] \times [0, 1] \to \mathbf{R}$ with $a(0, x) = a(1, x) = a(1, x) = a(x, 1) = 0$, $x \in [0, 1]$, and $\mathcal{C}_1$ be the set of continuous functions $\tilde{a} : [0, 1] \to \mathbf{R}$ with $a(0) = a(1) = 0$. Then we see that $a_{ii'}^{(2)}(t, \cdot, *) \in \mathcal{C}^2$, $1 \leqq i < i' \leqq N$, $a_{ij}^{(1)} \in \mathcal{C}^1$, $i = 1, \ldots, N$, $j = 0, 1, \ldots, M$, for $\nu - a.e.\ (t, \omega)$.

Also, let $\mathcal{M} = (\mathcal{C}_2)^{N(N-1)/2} \times (\mathcal{C}_1)^{N(1+M)} \times \mathbf{R}^{J_2}$. Then $\mathcal{M}$ is a Poilish space. Let $Y(t, \omega) = ((a_{ii'}^{(2)}(t, \cdot, *, \omega))_{1 \leqq i < i' \leqq N}, (a_{ij}^{(1)}(t, \cdot, \omega))_{i=1,\ldots,N, j=0, 1, \ldots, M}, (a_{jj'}^{(0)}(t, \omega))_{(j, j') \in J_2})$. Then we see that $Y(t, \omega) \in \mathcal{M}$ for $\nu - a.e.\ (t, \omega)$. Therefore under the probability measure ν on $[0, \infty) \times \Omega$, $((\xi_i(t, \omega))_{i=1,\ldots,N}, Y(t, \omega), \theta(t, \omega))$ is a $[0, \infty)^N \times \mathcal{M} \times \bar{\Theta}$-valued ranndom variable. Let Γ be the support of the probability law of this random variable.

Then we have the following.

Proposition 8. *If* $((\xi_i)_{i=1,\dots,N}, (\tilde{a}^{(2)}_{ii'})_{1\leqq i<i'\leqq N}, (\tilde{a}^{(1)}_{ij})_{i=1,\dots,N,j=0,1,\dots,M}, (\tilde{a}^{(0)}_{jj'})_{jj'})_{(j,j')\in J_2}, \theta)$ *belongs to* Γ, *and* $\theta \in \Theta$,*then*

$$\sum_{i=1}^{N} \xi_i (x_i \frac{\partial K}{\partial x_i}(x,\theta) - K(x,\theta))$$

$$+ \sum_{1\leqq i<i'\leqq N} \tilde{a}^{(2)}_{ii'}(x_i, x_{i'})(S^{(2)}_{ii'}K)(x,\theta) + \sum_{i=1}^{N}\sum_{j=0}^{d} \tilde{a}^{(1)}_{ij}(x_i)(S^{(1)}_{ij}K)(x,\theta)$$

$$+ \sum_{(j,j')\in J_2} \tilde{a}^{(0)}_{jj'}(S^{(0)}_{jj'}K)(x,\theta) = 0, \text{ for all } x = (x_1,\dots,x_N) \in (0,1)^N$$

and

$$\sum_{1\leqq i<i'\leqq N} \tilde{a}^{(2)}_{ii'}(x_i, x_{i'})z_i z_{i'} + \sum_{i=1}^{N} \tilde{a}^{(1)}_{i0}(x_i)z_i^2 + \sum_{i=1}^{N}\sum_{j=1}^{d} \tilde{a}^{(1)}_{ij}(x_i)z_i y_j$$

$$+ \sum_{(j,j')\in J_2} \tilde{a}^{(0)}_{jj'} y_j y_{j'} \geqq 0$$

for all $x = (x_1,\dots,x_N) \in (0,1)^N$ *and* $z_1,\dots,z_N, y_1,\dots,y_M \in \mathbf{R}$.

Let Γ_θ be the support of $\theta(t,\omega)$ under ν. Let $\pi : [0,\infty)^N \times \mathcal{M} \times \bar{\Theta} \to \bar{\Theta}$ be the natural projection, and let $\Gamma_0 = \pi(\Gamma \cap ((0,\infty)\times[0,\infty)^{N-1}\times\mathcal{M}\times\Theta))$. Then Proposition 7 implies that the closure of Γ_0 contains $\Gamma_\theta \cap \Theta$.

Then we have the following from the previous Proposition.

Lemma 9. *Let* $N \geqq 2$, $M \geqq 1$, Θ *be an open set in* $\mathbf{R}^M$, *and* $K \in C([0,1]^N \times \Theta; [0,1])$. *Assume that* K *is an admissible family of copula functions and that* $K|_{(0,1)^N\times\Theta}$ *is* C^2. *Then there is a subset* A *of* Θ *such that the closure of* A *contains non-void open set in* Θ, *and for any* $\theta \in A$, *there are* $\xi_1 > 0$, $\xi_i \geqq 0$, $i = 2,\dots,N$, $\tilde{a}^{(2)}_{ii'} \in \mathcal{C}_1$, $1 \leqq i < i' \leqq N$, $\tilde{a}^{(1)}_{ij} \in \mathcal{C}_1$, $i = 1,\dots,N$, $j = 0,1,\dots,M$, *and* $\tilde{a}^{(0)}_{jj'} \in \mathbf{R}$, $(j,j') \in J_2$, *such that*

$$\sum_{i=1}^{N} \xi_i (x_i \frac{\partial K}{\partial x_i}(x,\theta) - K(x,\theta))$$

$$+ \sum_{1\leqq i<i'\leqq N} \tilde{a}^{(2)}_{ii'}(x_i, x_{i'})(S^{(2)}_{ii'}K)(x,\theta) + \sum_{i=1}^{N}\sum_{j=0}^{d} \tilde{a}^{(1)}_{ij}(x_i)(S^{(1)}_{ij}K)(x,\theta)$$

$$+ \sum_{(j,j')\in J_2} \tilde{a}^{(0)}_{jj'}(S^{(0)}_{jj'}K)(x,\theta) = 0, \text{ for all } x = (x_1,\dots,x_N) \in (0,1)^N$$

and

$$\sum_{1\leqq i<i'\leqq N} \tilde{a}^{(2)}_{ii'}(x_i, x_{i'})z_i z_{i'} + \sum_{i=1}^{N} \tilde{a}^{(1)}_{i0}(x_i)z_i^2 + \sum_{i=1}^{N}\sum_{j=1}^{d} \tilde{a}^{(1)}_{ij}(x_i)z_i y_j$$
$$+ \sum_{(j,j')\in J_2} \tilde{a}^{(0)}_{jj'} y_j y_{j'} \geqq 0$$

for all $x = (x_1, \ldots, x_N) \in (0,1)^N$ *and* $z_1, \ldots, z_N, y_1, \ldots, y_M \in \mathbf{R}$.

5. Verification

Let $N, M \geqq 1$, and Θ be an open set in $\mathbf{R}^M$. Let $n \geqq 1$, and $\vec{z} = (z_{ik})_{i=1,\ldots,N,k=1,\ldots,n} \in (0,1)^{nN}$.

For $\vec{k} = (k_1, \ldots, k_N) \in \{1, \ldots, n\}^N$, and $\vec{z} \in (0,1)^{nN}$, let $Z_i(\vec{z}, \vec{k}) = z_{ik_i}$, $i = 1, \ldots, N$, and $\vec{Z}(\vec{z}, \vec{k}) = (z_{1k_1}, \ldots, z_{Nk_N}) \in (0,1)^N$.

Let $K \in C([0,1]^N \times \Theta; [0,1])$ be an admissible family of copula functions, and assume that $K|_{(0,1)^N\times\Theta}$ is C^2. Now let A be a subset in Θ as in Lemma 9 Then for any $\theta \in A$, there are ξ_i, $i = 1, \ldots, N$, $\tilde{a}^{(2)}_{ii'}$, $1 \leqq i < i' \leqq N$, $\tilde{a}^{(1)}_{ij}$, $i = 1, \ldots, N, j = 0, 1, \ldots, M$, and $\tilde{a}^{(0)}_{jj'}$, $(j, j') \in J_2$, be as in Lemma 9. Then we have

$$\sum_{i=1}^{N} \xi_i(Z_i(\vec{z},\vec{k})\frac{\partial K}{\partial x_i}(\vec{Z}(\vec{z},\vec{k}),\theta) - K(\vec{Z}(\vec{z},\vec{k}),\theta))$$
$$+ \sum_{1\leqq i<i'\leqq N} \tilde{a}^{(2)}_{ii'}(Z_i(\vec{z},\vec{k}), Z_{i'}(\vec{z},\vec{k}))(S^{(2)}_{ii'}K)(\vec{Z}(\vec{z},\vec{k}),\theta)$$
$$+ \sum_{i=1}^{N}\sum_{j=0}^{d} \tilde{a}^{(1)}_{ij}(Z_i(\vec{z},\vec{k}))(S^{(1)}_{ij}K)(\vec{Z}(\vec{z},\vec{k}),\theta)$$
$$+ \sum_{(j,j')\in J_2} \tilde{a}^{(0)}_{jj'}(S^{(0)}_{jj'}K)(\vec{Z}(\vec{z},\vec{k}),\theta) = 0.$$

That is

$$\sum_{i=1}^{N} \xi_i(Z_i(\vec{z},\vec{k})\frac{\partial K}{\partial x_i}(\vec{Z}(\vec{z},\vec{k}),\theta) - K(\vec{Z}(\vec{z},\vec{k}),\theta))$$
$$+ \sum_{p,q=1}^{n} \sum_{1\leqq i<i'\leqq N} \tilde{a}^{(2)}_{ii'}(z_{ip}, z_{i'q})\delta_{p,k_i}\delta_{q,k_{i'}}(S^{(2)}_{ii'}K)(\vec{Z}(\vec{z},\vec{k}),\theta)$$

$$+\sum_{p=1}^{n}\sum_{i=1}^{N}\sum_{j=0}^{d}\tilde{a}_{ij}^{(1)}(z_{ip})\delta_{p,k_i}(S_{ij}^{(1)}K)(\vec{Z}(\vec{z},\vec{k}),\theta)$$

$$+\sum_{(j,j')\in J_2}\tilde{a}_{jj'}^{(0)}(S_{jj'}^{(0)}K)(\vec{Z}(\vec{z},\vec{k}),\theta)=0, \qquad \vec{k}\in\{1,\ldots,n\}^N. \tag{6}$$

Let

$$C_n^{(2)}=\{(i,i')\in\{1,2,\ldots,N\}^2;\ i<i'\}\times\{1,2,\ldots,n\}^2$$

and

$$C_n^{(1)}=\{1,2,\ldots,N\}\times\{0,1,\ldots,M\}\times\{1,2,\ldots,n\}.$$

For any $G\in C^2((0,1)^N\times\Theta)$, $n\geqq 1$, and $\vec{k}\in\{1,\ldots,n\}^N$, we define continuous functions defined in $(0,1)^{nN}\times\Theta$, $(M_i^{(n)I}G)(\cdot,\vec{k})$, $i=1,\ldots,N$, $(M_{ii'pq}^{(n)(2)}G)(\cdot,\vec{k})$, $(i,i',p,q)\in C_n^{(2)}$, $(M_{ijp}^{(n)(1)}G)(\cdot,\vec{k})$, $(i,j,p)\in C_n^{(1)}$, $(M_{jj'}^{(n)(0)}G)(\cdot,\vec{k})$, $(j,j')\in J_2$, by

$$(M_i^{(n)I}G)(\vec{z},\theta,\vec{k})=Z_i(\vec{z},\vec{k})\frac{\partial G}{\partial x_i}(\vec{Z}(\vec{z},\vec{k}),\theta)-G(\vec{Z}(\vec{z},\vec{k}),\theta)\ i=1,\ldots,N,$$

$$(M_{ii'pq}^{(n)(2)}G)(\vec{z},\theta,\vec{k})=\delta_{p,k_i}\delta_{q,k_{i'}}(S_{ii'}^{(2)}G)(\vec{Z}(\vec{z},\vec{k}),\theta)\quad (i,i',p,q)\in C_n^{(2)},$$

$$(M_{ijp}^{(n)(1)}G)(\vec{z},\theta,\vec{k})=\delta_{p,k_i}(S_{ij}^{(1)}G)(\vec{Z}(\vec{z},\vec{k}),\theta),\quad (i,j,p)\in C_n^{(1)},$$

$$(M_{jj'}^{(n)(0)}G)(\vec{z},\theta,\vec{k})=(S_{jj'}^{(0)}G)(\vec{Z}(\vec{z},\vec{k}),\theta),\quad (j,j')\in J_2,$$

for any $\vec{z}\in(0,1)^{nN}$ and $\theta\in\Theta$.

Let $C_{n0}=C_n^{(2)}\cup C_n^{(1)}\cup J_2$, and $C_n=\{1,\ldots,N\}\cup C_{n0}$. Note that the cardinal $\#(C_{n0})$ of C_{n0} is equal to $n^2N(N-1)/2+nN(M+1)+M(M+3)/2$, and $\#(C_n)=N+\#(C_{n0})$.

For any $G\in C^2((0,1)^N\times\Theta)$, $n\geqq 1$, and $\gamma\in C_n$ we define a continuous function $(\vec{M}^{(n)}G)_\gamma:(0,1)^{nN}\times\Theta\to\mathbf{R}^{\{1,\ldots,n\}^N}$ by

$$(\vec{M}^{(n)}G)_\gamma(\vec{z},\theta)$$

$$=\begin{cases}((M_{ii'pq}^{(n)(2)}G)(\vec{z},\theta,\vec{k}))_{\vec{k}\in\{1,\ldots,n\}^N} & \text{if } \gamma=(i,i',p,q)\in C_n^{(2)},\\ (M_{ijp}^{(n)(1)}G)(\vec{z},\theta,\vec{k}))_{\vec{k}\in\{1,\ldots,n\}^N} & \text{if } \gamma=(i,j,p)\in C_n^{(1)},\\ (M_{jj'}^{(n)(0)}G)(\vec{z},\theta,\vec{k}))_{\vec{k}\in\{1,\ldots,n\}^N} & \text{if } \gamma=(j,j')\in J_2,\\ ((M_i^{(n)I}G))(\vec{z},\theta,\vec{k}))_{\vec{k}\in\{1,\ldots,n\}^N} & \text{if } \gamma=i\in\{1,\ldots,N\}.\end{cases}$$

For any $G \in C^2((0,1)^N \times \Theta)$, $n \geqq 1$, $\vec{z} \in (0,1)^{nN}$ and $\theta \in \Theta$, let $V_n(G, \vec{z}, \theta)$ (resp. $V_{n0}(G, \vec{z}, \theta)$) be the vector subspace of $\mathbf{R}^{\{1,\dots,n\}^N}$ spaned by $\{(\vec{M}^{(n)}G)_\gamma(\vec{z}, \theta);\ \gamma \in C_n\}$ (resp. $\{(\vec{M}^{(n)}G)_\gamma(\vec{z}, \theta);\ \gamma \in C_{n0}\}$). Also, let $N_{(n)}(G, \vec{z}, \theta)$ be a vector space in $\mathbf{R}^N$ given by

$$N_{(n)}(G, \vec{z}, \theta) = \{(v_1, \dots, v_N) \in \mathbf{R}^N; \sum_{i=1}^n v_i (\vec{M}^{(n)}G)_i(\vec{z}, \theta) \in V_{n0}(G, \vec{z}, \theta)\}.$$

Then we have $N_{(n)}(K, \vec{z}, \theta) \cap [0, \infty)^N \neq \{0\}$, for any $\theta \in A$.

Therefore we have the following.

Lemma 10. *Let $N \geqq 2$, $M \geqq 1$, and Θ be an open subset of $\mathbf{R}^M$ Let $K \in C([0,1]^N \times \Theta; [0,1])$. Assume that K is an admissible family of copula functions, and that $K|_{(0,1)^N \times \Theta}$ is C^2. Then there is a subset A of Θ such that the closure of A contains non-void open set in Θ, and for any $\theta \in A$ and $\vec{z} \in (0,1)^{nN}$, $N_{(n)}(K, \vec{z}, \theta) \cap [0, \infty)^N \neq \{0\}$.*

As a corollary we have the following.

Corollary 11. *Let $N \geqq 2$, $M \geqq 1$, and Θ be an open subset of $\mathbf{R}^M$. Let $K \in C([0,1]^N \times \Theta; [0,1])$. Assume that K is an admissible family of copula functions, and that $K|_{(0,1)^N \times \Theta}$ is C^2. Then for any $n \geqq 1$ and $\vec{z} \in (0,1)^{nN}$, there is a non-void open subset U of Θ such that*

$$\dim V_n(K, \vec{z}, \theta) \leqq \#(C_n) - 1, \qquad \theta \in U.$$

Proof. Since $\dim V_n(K, \vec{z}, \theta) \leqq n^N$, the assertion is obvious in the case that $n^N \leqq \#(C_n) - 1$. So assume that $n^N \geqq \#(C_n)$.

Let A be a subset in Θ as in Lemma 10. It is easy to see that for any $\theta \in A$

$$\dim V_n(K, \vec{z}, \theta) = \dim V_{n0}(K, \vec{z}, \theta) + N - \dim N_{(n)}(G, \vec{z}, \theta) \leqq \#(C_n) - 1.$$

Let H be the set of injections from C_n to $\{1, \dots, n\}^N$ and let

$$\varphi(\theta) = \sum_{h \in H} \det((\vec{M}^{(n)}G)_{\gamma_1}(\vec{z}, \theta, h(\gamma_2)))_{\gamma_1, \gamma_2 \in C_n})^2, \quad \theta \in \Theta.$$

Then we see $\varphi(\theta) = 0$ for $\theta \in A$. Since $\varphi : \theta \to \mathbf{R}$ is continuous, we see that $\varphi(\theta) = 0$ for $\theta \in \bar{A}$. So we see that

$$\dim V_n(K, \vec{z}, \theta) \leqq \#(C_n) - 1, \qquad \theta \in \bar{A}.$$

This implies our assertion. ■

6. Proof of Theorem 2

Now let $N \geqq 2$, $M \geqq 1$ and $n \geqq 1$. We say that $h : C_n^{(2)} \cup C_n^{(1)} \to \{1, \dots, n\}^N$ is a matching map, if h is injective and satisfying the following.

$$h((i, i', p, q))_i = p, \quad h((i, i', p, q))_{i'} = q \quad \text{for any } (i, i', p, q) \in C_n^{(2)},$$

and

$$h((i, j, p))_i = p \quad \text{for any } (i, j, p) \in C_n^{(1)}.$$

Proposition 12. *Let $N \geqq 3$. Assume that there is a matching map $h_0 : C_n^{(2)} \cup C_n^{(1)} \to \{1, \dots, n\}^N$, and that $\#(C_n) \leqq n^N$. Let $0 < c_{i1} < c_{i2} < \dots < c_{in} < 1$, $i = 1, \dots, N$, $\vec{c} = (c_{ik})_{i=1,\dots,N,k=1,\dots,n} \in (0, 1)^{nN}$, and $\theta_0 \in \mathbf{R}^M$. Then there is a $K \in \mathcal{C}_{(N)}(\mathbf{R}^M)$ such that* $\dim V_n(K, \vec{c}, \theta_0) = \#(C_n)$.

Proof. From the assumption, there is an injective map $h : C_n \to \{1, \dots, n\}^N$ such that the restriction of h to $C_n^{(2)} \cup C_n^{(1)}$ is equal to h_0. Note that $\vec{Z}(\vec{c}, \vec{k})$, $\vec{k} \in \{1, \dots, n\}^N$, are distinct points. Let

$$\varepsilon_0 = \min\{|\vec{Z}(\vec{c}, \vec{k}) - \vec{Z}(\vec{c}, \vec{k}')|;\ \vec{k}, \vec{k}' \in \{1, \dots, n\}^N, \vec{k} \neq \vec{k}'\},$$

$$\varepsilon_1 = \min\{c_{i1};\ i = 1, \dots, N\} \wedge \min\{1 - c_{in};\ i = 1, \dots, N\},$$

and $\varepsilon = \varepsilon_0 \wedge \varepsilon_1$. Let $\varphi_0 \in C_0^\infty(\mathbf{R}^N)$ and $\varphi_1 \in C_0^\infty(\mathbf{R}^M)$ such that $\varphi_0(x) = 1$, $|x| < \varepsilon/3$, $\varphi_0(x) = 0$, $|x| > 2\varepsilon/3$, and $\varphi_1(\theta) = 1$, $|\theta| < 1$.

Let $F : \mathbf{R}^N \times \mathbf{R}^M \times C_n \to \mathbf{R}$ be given by the following.

$$F(x, \theta, i) = -\varphi_0(x - \vec{Z}(\vec{c}, h(i))), \quad i \in I,$$

$$F(x, \theta, (0, j)) = \varphi_0(x - \vec{Z}(\vec{c}, h((0, j)))(\theta_j - \theta_{0j})\varphi_1(\theta - \theta_0),$$
$$j = 1, \dots, M,$$

$$F(x, \theta, (j, j)) = \frac{1}{2}\varphi_0(x - \vec{Z}(\vec{c}, h((j, j)))(\theta_j - \theta_{0j})^2\varphi_1(\theta - \theta_0),$$
$$j = 1, \dots, M,$$

$$F(x, \theta, (j, j')) = \varphi_0(x - \vec{Z}(\vec{c}, h((j, j')))(\theta_j - \theta_{0j})(\theta_{j'} - \theta_{0j'})\varphi_1(\theta - \theta_0),$$
$$1 \leqq j < j' \leqq M,$$

$$F(x, \theta, (i, i', p, q)) = \varphi_0(x - \vec{Z}(\vec{c}, h((i, i', p, q)))(x_i - \vec{Z}(\vec{c}, h((i, i', p, q))_i)$$
$$(x_{i'} - \vec{Z}(\vec{c}, h((i, i', p, q))_{i'}),$$

$(i, i', p, q) \in C_n^{(2)}$,

$$F(x, \theta, (i, 0, p)) = \frac{1}{2}\varphi_0(x - \vec{Z}(\vec{c}, h((i, 0, p)))(x_i - \vec{Z}(\vec{c}, h((i, 0, p))_i)^2,$$

$$(i, 0, p) \in C_n^{(1)},$$

$$F(x, \theta, (i, j, p))$$
$$= \varphi_0(x - \vec{Z}(\vec{c}, h((i, j, p)))(x_i - \vec{Z}(\vec{c}, h((i, j, p, q))_i)(\theta_j - \theta_{0j})\varphi_1(\theta - \theta_0),$$
$$(i, j, p) \in C_n^{(1)}, j \geqq 1.$$

Then we see that $(\vec{M}^{(n)} F(\cdot, u))_\gamma(\vec{c}, \theta_0, h(\alpha)) = \delta_{\gamma\alpha}$, $\gamma, \alpha \in C_n$, and that $(\vec{M}^{(n)} F(\cdot, u))_\gamma(\vec{c}, \theta_0, \vec{k}) = 0$, $\gamma \in C_n$, $\vec{k} \in \{1, \ldots, n\}^N$, with $\vec{k} \neq h(\gamma)$. Now let $F_0 \in C_0^\infty(\mathbf{R}^N \times \mathbf{R}^M)$ be given by

$$F_0(\cdot, *) = \sum_{\gamma \in C_n} F(\cdot, *, \gamma).$$

Then we see that derivatives of F_0 of any order are bounded functions defined in $\mathbf{R}^N \times \mathbf{R}^M$,

$$\det((\vec{M}^{(n)} F_0)_\gamma(\vec{c}, \theta_0, h(\alpha))_{\gamma,\alpha\in C_n}) = 1,$$

and $F_0(x_1, \ldots, x_N, \theta) = 0$, if $x_i < \varepsilon/3$ or $x_i > 1 - (\varepsilon/3)$ for some $i = 1, \ldots, N$.

Let

$$f_0(x, \theta) = \frac{\partial^N F_0}{\partial x_1 \ldots \partial x_N}(x, \theta)$$

Then we have

$$F_0(x, \theta) = \int_0^{x_1} \cdots \int_0^{x_N} f_0(y_1, \ldots, y_N, \theta) dy_1 \ldots dy_N.$$

Let $c = \sup\{|f_0(x, \theta)|;\ (x, \theta) \in \mathbf{R}^N \times \mathbf{R}^M\} < \infty$, and let

$$G_s(x, \theta) = \int_0^{x_1} \cdots \int_0^{x_N} (1 + s f_0(y_1, \ldots, y_N, \theta)) dy_1 \ldots dy_N$$
$$= x_1 \cdots x_N + s F_0(x_1, \ldots, x_N.\theta)$$

for $s \in \mathbf{R}$, $x \in [0, 1]^N$, $\theta \in \mathbf{R}^M$. Then

$$p(s) = \det(((\vec{M}^{(n)} G_s)_\gamma(\vec{c}, \theta_0, h(\alpha)))_{\gamma,\alpha\in C_n})$$

is a polynomial in s and

$$\lim_{s\to\infty} s^{-\#(C_n)} p(s) = 1.$$

Therefore there is a $\tilde{s}$ with $0 < \tilde{s} < 1/(2c+1)$ such that

$$\det(((\vec{M}^{(n)} G_{\tilde{s}})_\gamma(\vec{c}, \theta_0, h(\alpha)))_{\gamma,\alpha\in C_n}) \neq 0.$$

Note that $1 + \tilde{s} f_0(y_1, \ldots, y_N, \theta) > 1/(2c+1)$. So it is easy to see that $G_{\tilde{s}}(\cdots, \theta)$ is a copula function for all $\theta \in \mathbf{R}^M$.

This shows our assertion. ■

Proposition 13. *Let $N = 3$, $M \geqq 1$ and $n \geqq 1$. Suppose that $n \geqq M + 3$, and $n \equiv 1$ or $5 \bmod 6$. Then there is a matching map $h : C_n^{(2)} \cup C_n^{(1)} \to \{1, \ldots, n\}^N$.*

Proof. First we prove the following.

Claim. Let $p, q, r = 1, \ldots, n$. If $p \neq q$ and $r \equiv 2q - p \bmod n$, then $r \neq p, q$, $p \not\equiv 2r - q \bmod n$, and $q \not\equiv 2p - r \bmod n$.

Actually if $r = p$, we have $2p \equiv 2q \bmod n$, which implies $p = q$. If $r = q$, we have $q \equiv p \bmod n$. which implies $p = q$. If $p \equiv 2r - q \bmod n$, then $3p \equiv 3q \bmod n$, which implies $p = q$. If $q \equiv 2p - r \bmod n$, then $3q \equiv 3p \bmod n$, which implies $p = q$. Therefore we have our Claim.

Let us define $h : C_n^{(2)} \cup C_n^{(1)} \to \{1, \ldots, n\}^N$ by the following.

$h|_{C_n^{(2)}}$ is given by the following. For $p, q = 1, \ldots, n$, with $p \neq q$

$$h(1,2,p,q) = (p,q,r), \ \ h(1,3,q,p) = (q,r,p) \ \ h(2,3,p,q) = (r,p,q),$$

where $r = 1, \ldots, n$, with $r \equiv 2q - p \bmod n$. For $p = 1, \ldots, n$,

$$h(1,2,p,p) = (p,p,r), \ h(1,3,p,p) = (p,r,p) \ h(2,3,p,p) = (r,p,p),$$

where $r = 1, \ldots, n$, with $r \equiv p - 1 \bmod n$.

$h|_{C_n^{(1)}}$ is given by the following. For $p = 1, \ldots, n$, and $j = 0, 1, \ldots, M$,

$$h(1,j,p) = (p,r,r), \quad h(2,j,p) = (r,p,r) \quad h(3,j,p) = (r,r,p),$$

where $r = 1, \ldots, n$, with $r \equiv p + j + 2 \bmod n$.

By the above Claim, we can easily check h is a matching map. ■

Proposition 14. *Let $N \geqq 4$, $M \geqq 1$ and $n \geqq 1$. Suppose that $n \geqq M + 2$. Then there is a matching map $h : C_n^{(2)} \cup C_n^{(1)} \to \{1, \ldots, n\}^N$.*

Proof. First take a map $R : \{i, \ldots, n\}^2 \to \{1, \ldots, n\}$ such that $R(p,q) \neq p, q$ and $R(p,p) \equiv p + 1 \bmod n$. Since $n \geqq 3$, we can take such a map. Now let us define $h : C_n^{(2)} \cup C_n^{(1)} \to \{1, \ldots, n\}^N$ by the following.

$h(i, i', p, q) = (k_1, \ldots, k_N)$, $(i, i', p, q) \in C_n^{(2)}$, where $k_i = p$, $k_{i'} = q$, $k_r = R(p,q)$, $r \neq i, i'$. $h(i, j, p) = (k_1, \ldots, k_N)$, $(i, j, p) \in C_n^{(1)}$, where $k_i = p$, $k_r = 1, \ldots, n$, with $k_r \equiv p + j + 1 \bmod n$.

Since $p + 2 \not\equiv p \bmod n$, we can easily check h is a matching map. ■

Now let us prove Theorem 2.

Let $N \geqq 3$, $M \geqq 1$, and Θ is a non-empty open set in $\mathbf{R}$. By Propositions 13 and 14, there are $n \geqq 1$ and an injective map $h : C_n \to \{1, \ldots, n\}^N$ for which $h|_{C_n^{(2)} \cup C_n^{(1)}}$ is a matching map. Fix such an n and let us take $\vec{c} \in \mathbf{R}^{Nn}$ such that $0 < c_{i1} < c_{i2} < \ldots < c_{in} < 1$, $i = 1, \ldots, N$, $i = 1, \ldots, N$. Let $D(\theta)$, $\theta \in \Theta$, be a set given by

$$D(\theta) = \{K \in \mathcal{C}_{(N)}(\Theta);\ \dim V_n(K, \vec{c}, \theta_0) = \#(C_n)\}.$$

Then by Propositions 12, we see that $D(\theta) \neq$ for all $\theta \in \Theta$. Let H be the set of injections from C_n to $\{1, \ldots, n\}^N$. Let $\varphi : \mathcal{C}_{(N)}(\Theta) \to \mathbf{R}$ be given by

$$\varphi(K) = \sum_{h \in H} \det((\vec{M}^{(n)} K)_{\gamma_1}(\vec{z}, \theta, h(\gamma_2)))_{\gamma_1, \gamma_2 \in C_n})^2.$$

Then we see that $\varphi : \mathcal{C}_{(N)}(\Theta) \to \mathbf{R}$ is continuous and $D(\theta) = \{K \in \mathcal{C}_{(N)}(\Theta);\ \varphi(K) > 0\}$. So we see that $D(\theta)$ is an open subset $\mathcal{C}_{(N)}(\Theta)$.

Let $G \in D(\theta)$. For any $K \in \mathcal{C}_{(N)}(\Theta)$ and $s \in [0, 1]$, $(1 - s)K + sG_0 \in \mathcal{C}_{(N)}(\Theta)$. Also, $\varphi((1 - s)K + sG)$ is a polynomial in s, and so is not equal to 0 except finite s's. Therefore there is a $\{s_\ell\}_{\ell=1}^{\infty} \subset [0, 1]$ such that $s_\ell \downarrow 0$, $\ell \to \infty$, and $(1 - s_\ell)K + s_\ell)G) \in D(\theta)$, $\ell \geqq 1$. This observation shows that $D(\theta)$ is dense in $\mathcal{C}_{(N)}(\Theta)$ for all $\theta \in \Theta$.

Now let $\{\theta_m\}_{m=1}^{\infty}$ be a dense set in Θ, and let

$$D = \bigcap_{m=1}^{\infty} D(\theta_m).$$

Then by Corollary 11, we see that any element of D is not admissible family of copula functions. This proves Theorem 2.

7. Remarks

Let $N \geqq 3$, $M \geqq 1$, and Θ be an open set in $\mathbf{R}^M$. Let $K \in \mathcal{C}_{(N)}(\Theta)$. Assume that $n \geqq N$.

Let

$$\begin{aligned} &A^{(2)} \\ &= \{(1, 2, p, q);\ p, q = 1, \ldots, n\} \cup \{(1, i, p, q);\ i = 3, \ldots, N, \\ &\qquad p = 1, \ldots, n,\ q = 2, \ldots, n\} \\ &\qquad \cup \{(i, i', p, q);\ 2 \leqq i < i' \leqq N,\ p, q = 2, \ldots, n\} \subset C_n^{(2)}. \end{aligned}$$

Also, let $\vec{k}_{ii'pq} \in \mathbf{R}^{\{1,\ldots,n\}^N}$, $(i,i',p,q) \in C_n^{(2)}$, be given by

$$\vec{k}_{ii'pq} = (1,\ldots,1,\underset{i}{p},1,\ldots,1,\underset{i'}{q},1,\ldots,1).$$

Then we have the following.

Proposition 15. *Let* $\theta_0 \in \Theta$, *and assume that*

$$\frac{\partial^2 K}{\partial x_i \partial x_{i'}}(x,\theta_0) > 0, \qquad x \in (0,1)^N,\ 1 \leqq i < i' \leqq N.$$

Then for any $n \geqq N$, *and* $\vec{z} \in (0,1)^{3n}$,

$$\dim V_{n0}(K,\vec{c},\theta_0) \geqq \#(A^{(2)}) = \frac{N(N-1)}{2}n^2 - nN(N-2) + \frac{(N-1)(N-2)}{2}.$$

Proof. Remind that

$$(M^{(n)(2)}_{ii'pq}K)(\vec{z},\theta_0,\vec{k}) = \delta_{p,k_i}\delta_{q,k_{i'}}\frac{\partial^2 K}{\partial x_i \partial x_{i'}}(z_{1k_1},\ldots,z_{Nk_N},\theta_0), \quad p,q = 1,\ldots,n.$$

So for $(i,i',p,q),(j,j',r,\ell) \in A^{(2)}$, we see that $(M^{(n)(2)}_{ii'pq}K)(\vec{z},\theta_0,\vec{k}_{jj'r\ell}) = 0$ if $i > j$, or if $i = j$ and $i' > j$, and that $(M^{(n)(2)}_{ii'pq}K)(\vec{z},\theta_0,\vec{k}_{ii'r\ell}) = \delta_{p,r}\delta_{q,\ell}c_{ii'pq}$, for some positive numbers $c_{ii'pq}$. So we see that $\{(\vec{M}^{(n)}K)_\gamma(\vec{z},\theta_0);\ \gamma \in A^{(2)}\}$ is linearly independent. So we have our assertion. ■

From now on we think of a special case. We assume that K is a family of Archimedian copula functions, i.e., there are smooth functions $\varphi : (0,1)\times\Theta \to (0,\infty)$ and $\rho : (0,\infty)\times\Theta \to (0,1)$ such that

$$K(x_1,\ldots,x_N,\theta) = \rho(\sum_{k=1}^N \varphi(x_k,\theta),\theta), \qquad x_1,\ldots,x_N \in (0,1),\ \theta \in \Theta.$$

Then $\rho(\cdot,\theta)$ must be the inverse function of $\varphi(\cdot,\theta)$.

Then we have the following.

Proposition 16. *Let*

$$m_0 = \frac{N(N-1)}{2}n^2 + N(M+3-N)n - \frac{(N-1)(2M+4-N)}{2} + \frac{M(M+3)}{2}.$$

Then we have the following.

(1) $\dim V_{n0}(K,\vec{z},\theta) \leqq m_0$. *and* $\dim V_n(K,\vec{z},\theta) \leqq m_0 + 1$ *for any* $\vec{z} \in (0,1)^{Nn}$, *and* $\theta \in \Theta$.

(2) *Assume that* Θ *is connected and that* $\varphi : (0,1) \times \Theta \to (0,\infty)$ *is real analytic. If there exists a* $\theta_0 \in \Theta$ *such that* $\dim V_n(K, \vec{z}, \theta_0) = m_0 + 1$, *then* K *is not an admissible family of copula functions.*

Since the proof is rather long, we will give it in the next section.

Now let us think of a family of Gumbel copula functions. Let $N = 3$, $M = 1$, and $\Theta = (0, 1)$. Let $K \in \mathcal{C}_{(3)}((0, 1))$ be given by

$$K(x_1, x_2, x_3, \theta) = \exp(-(\sum_{i=1}^{3}(-\log x_i)^{\theta})^{1/\theta}), \qquad x_1, x_2, x_3 \in (0, 1),\ \theta \in (0, 1).$$

Then letting $\varphi(x, \theta) = (-\log x)^{\theta}$, $\rho(y, \theta) = \exp(-y^{1/\theta})$, we see that

$$K(x_1, x_2, x_3, \theta) = \rho(\sum_{i=1}^{3}\varphi(x_i, \theta), \theta), \qquad x_1, x_2, x_3 \in (0, 1),\ \theta \in (0, 1).$$

Let $n = 5$. Then we have $m_0 = 89$. So by Proposition 16 we see that if there exist $\theta_0 \in (0, 1)$ and $\vec{z} = (z_{ip})_{i=1,2,3,p=1,\ldots,5} \in (0, 1)^{15}$ such that $\dim V_n(K, \vec{z}, \theta_0) = 90$, we see that K is not admissible family of copula functions.

By using numerical computation, we check that $\dim V_n(K, \vec{z}, \theta_0) = 90$ for $(z_{i1}, \ldots, z_{i5}) = (0.55, 0.65, 0.75, 0.85, 0.95)$, $i = 1, 2, 3$, and $\theta_0 = 0.4$ or 0.6. Actuary, we compute the dimension of the vector subspace in $\mathbf{R}^{\{1,2,3,4,5\}^3}$ spanned by $e^{(2)}_{ii'pq}(\vec{z}, \theta_0)$, $(i, i', p, q) \in A^{(2)}$, $e^{(1)}_{ijp}(\vec{z}, \theta_0)$, $(i, j, p) \in A^{(1)}$, $e^{(0)}_{jj'}(\vec{z}, \theta_0)$, $(j, j') \in J_2$, and $e_0(\vec{z}, \theta_0)$ given in the next section by applying Householder transformation for the associated matrix, and we are convinced that it is 90. As we show in the next section, the dimension of this vector subspace is the same as $\dim V_n(K, \vec{z}, \theta_0)$ in this case.

8. Proof of Proposition 16

For $\vec{z} \in (0, 1)^{3n}$, $\theta \in \Theta$ and $\vec{k} \in \{1, \ldots, n\}^3$, let

$$\Phi(\vec{z}, \theta, \vec{k}) = \sum_{i=1}^{3}\varphi(z_{ik_i}, \theta),$$

$$e^{(2)}_{ii'pq}(\vec{z}, \theta, \vec{k}) = \delta_{pk_i}\delta_{qk_{i'}}\frac{\partial^2 \rho}{\partial y^2}(\Phi(\vec{z}, \theta, \vec{k}), \theta), \quad (i, i', p, q) \in C^{(2)}_n,$$

$$e_{i0p}^{(1)}(\vec{z},\theta,\vec{k}) = \delta_{pk_i}\frac{\partial \rho}{\partial y}(\Phi(\vec{z},\theta,\vec{k}),\theta), \quad (i,0,p) \in C_n^{(1)},$$

$$e_{ijp}^{(1)}(\vec{z},\theta,\vec{k}) = \delta_{pk_i}\frac{\partial^2 \rho}{\partial \theta_j \partial y}(\Phi(\vec{z},\theta,\vec{k}),\theta) \quad (i,j,p) \in C_n^{(1)}, \ j \geqq 1,$$

$$e_{jj'}^{(0)}(\vec{z},\theta,\vec{k}) = \frac{\partial^2 \rho}{\partial \theta_j \partial \theta_{j'}}(\Phi(\vec{z},\theta,\vec{k}),\theta), \quad (j,j') \in J_2, \ j \geqq 1,$$

$$e_{0j}^{(0)}(\vec{z},\theta,\vec{k}) = \frac{\partial \rho}{\partial \theta_j}(\Phi(\vec{z},\theta,\vec{k}),\theta). \quad (0,j) \in J_2,$$

and

$$e_0(\vec{z},\theta,\vec{k}) = -\rho(\Phi(\vec{z},\theta,\vec{k}),\theta).$$

Let $e_{ii'pq}^{(2)}(\vec{z},\theta)$, $(i,i',p,q) \in C_n^{(2)}$, $e_{ijp}^{(1)}(\vec{z},\theta)$, $(i,j,p) \in C_n^{(1)}$, $e_{jj'}^{(0)}(\vec{z},\theta)$, $(j,j') \in J_2$, and $e_0(\vec{z},\theta)$ be elements of $\mathbf{R}^{\{1,\ldots,n\}^3}$ given by $e_{ii'pq}^{(2)}(\vec{z},\theta) = (e_{ii'pq}^{(2)}(\vec{z},\theta,\vec{k}))_{\vec{k}\in\{1,\ldots,n\}^3}$ etc.

Let $A^{(1)}$ be the subset of $C_n^{(1)}$ given by

$$\begin{aligned} &A^{(1)} \\ &= \{(1,j,p);\ j=0,1,\ldots,M,\ p=1,\ldots,n\} \\ &\quad \cup \{(i,j,p);\ i=2,\ldots,N,\ j=0,1,\ldots,M,\ p=2,\ldots,n\}, \end{aligned}$$

and $A^{(2)}$ be the subset of $C_n^{(2)}$ given in the previous section.

Let $U_0(\vec{z},\theta)$ be a vector subspace in $\mathbf{R}^{\{1,\ldots,n\}^N}$ spaned by $\{e_{ii'pq}^{(2)}(\vec{z},\theta);\ (i,i',p,q) \in A^{(2)}\}$, $\{e_{ijp}^{(1)}(\vec{z},\theta);\ (i,j,p) \in A^{(1)}\}$ and $\{e_{jj'}^{(0)}(\vec{z},\theta);\ (j,j') \in J_2\}$. Since $\#(A^{(2)}) + \#(A^{(1)}) + \#(J_2) = m_0$, we see that $\dim U_0(\vec{z},\theta) \leqq m_0$.

First, we prove the following.

Proposition 17. (1) $e_{ii'pq}^{(2)}(\vec{z},\theta) \in U_0(\vec{z},\theta)$ *for all* $(i,i',p,q) \in C_n^{(2)}$.

(2) $e_{ijp}^{(1)}(\vec{z},\theta) \in U_0(\vec{z},\theta)$ *for all* $(i,j,p) \in C_n^{(1)}$.

Proof. Let

$$\tilde{e}_{1p}^{(2)}(\vec{z},\theta) = \sum_{q=1}^{n} e_{12pq}^{(2)}(\vec{z},\theta) \in U_0(\vec{z},\theta), \quad p=1,\ldots,n,$$

and

$$\tilde{e}_{ip}^{(2)}(\vec{z},\theta) = \sum_{q=1}^{n} e_{1iqp}^{(2)}(\vec{z},\theta) \in U_0(\vec{z},\theta), \quad i=2,\ldots,N-1,\ p=2,\ldots,n.$$

Then we see that

$$e^{(2)}_{1ip1}(\vec{z},\theta) = \hat{e}^{(2)}_{1p}(\vec{z},\theta) - \sum_{q=2}^{n} e^{(2)}_{1ipq}(\vec{z},\theta) \in U_0(\vec{z},\theta),$$
$$i = 3,\ldots,N,\ p = 1,\ldots,n.$$

So we see that $e^{(2)}_{1ipq}(\vec{z},\theta) \in U_0(\vec{z},\theta)$, $i = 2,\ldots,N$, $p,q = 1,\ldots,n$.

Also, we see taht

$$e^{(2)}_{ii'1q}(\vec{z},\theta) = \hat{e}^{(2)}_{i'q}(\vec{z},\theta) - \sum_{p=2}^{n} e^{(2)}_{ii'pq}(\vec{z},\theta) \in U_0(\vec{z},\theta),$$
$$i = 2,\ldots,N,\ q = 2,\ldots,n,$$

and so

$$e^{(2)}_{ii'11}(\vec{z},\theta) = \hat{e}^{(2)}_{i1}(\vec{z},\theta) - \sum_{q=2}^{n} e^{(2)}_{ii'1q}(\vec{z},\theta) \in U_0(\vec{z},\theta), \quad i = 2,\ldots,N.$$

These show that the assetion (1).

Let

$$\tilde{e}^{(1)}_{j}(\vec{z},\theta) = \sum_{p=1}^{n} e^{(1)}_{1jp}(\vec{z},\theta) \in U_0(\vec{z},\theta), \quad j = 0,\ldots,M.$$

Then we see that

$$e^{(1)}_{ij1}(\vec{z},\theta) = \hat{e}^{(1)}_{i}(\vec{z},\theta) - \sum_{p=2}^{n} e^{(1)}_{ijp}(\vec{z},\theta) \in U_0(\vec{z},\theta), \quad j = 0,\ldots,M.$$

This proves the assertion (2). ■

Now note that

$$(M^{(n)(2)}_{ii'pq}K)(\vec{z},\theta,\vec{k}) = \frac{\partial\varphi}{\partial x}(z_{ip},\theta)\frac{\partial\varphi}{\partial x}(z_{i'q},\theta)e^{(2)}_{ii'pq}(\vec{z},\theta,\vec{k}),$$
$$(i,i',p,q) \in C^{(2)}_n,$$

$$(M^{(n)(1)}_{i0p}K)(\vec{z},\theta,\vec{k}) = \frac{\partial^2\varphi}{\partial x^2}(z_{ip},\theta)e^{(1)}_{i0p}(\vec{z},\theta,\vec{k})$$
$$+(\frac{\partial\varphi}{\partial x}(z_{ip},\theta))^2\tilde{e}^{(2)}_{ip}(\vec{z},\theta,\vec{k})), \quad (i,0,p) \in C^{(1)}_n,$$

$$(M_{ijp}^{(n)(1)}K)(\vec{z},\theta,\vec{k}) = \frac{\partial\varphi}{\partial x}(z_{ip},\theta)e_{ijp}^{(1)}(\vec{z},\theta,\vec{k})$$

$$+\frac{\partial\varphi}{\partial x}(z_{ip},\theta)(\sum_{q=1}^{n}\frac{\partial\varphi}{\partial\theta_j}(z_{iq},\theta)e_{iq}^{(2)}(\vec{z},\theta,\vec{k})) + \frac{\partial^2\varphi}{\partial\theta_j\partial x}(z_{ip},\theta)e_{i0p}^{(1)}(\vec{z},\theta,\vec{k}),$$

$$(i,j,p) \in C_n^{(1)},\ j \geqq 1,$$

$$(M_{0j}^{(n)(0)}K)(\vec{z},\theta,\vec{k}) = \sum_{i=1}^{N}\sum_{p=1}^{n}\frac{\partial\varphi}{\partial\theta}(z_{ip},\theta)e_{i0p}^{(1)}(\vec{z},\theta,\vec{k}) + e_{0j}^{(0)}(\vec{z},\theta,\vec{k}),$$

$$(0,j) \in J_2,$$

$$(M_{jj'}^{(n)(0)}K)(\vec{z},\theta,\vec{k})$$

$$= \sum_{1\leqq i<i'\leqq N}\sum_{p,q=1}^{n}(\frac{\partial\varphi}{\partial\theta_j}(z_{ip},\theta)\frac{\partial\varphi}{\partial\theta_{j'}}(z_{i'q},\theta)$$

$$+\frac{\partial\varphi}{\partial\theta_j}(z_{ip},\theta)\frac{\partial\varphi}{\partial\theta_{j'}}(z_{i'q},\theta))e_{ii'pq}^{(2)}(\vec{z},\theta,\vec{k})$$

$$+\sum_{i=1}^{N}\sum_{p=1}^{n}\frac{\partial\varphi_j}{\partial\theta_j}(z_{ip},\theta)\frac{\partial\varphi_{j'}}{\partial\theta_{j'}}(z_{ip},\theta)\tilde{e}_{ip}^{(2)}(\vec{z},\theta,\vec{k})$$

$$+\sum_{i=1}^{N}\sum_{p=1}^{n}\frac{\partial\varphi}{\partial\theta_j}(z_{ip},\theta)e_{ij'p}^{(1)}(\vec{z},\theta,\vec{k}) + \frac{\partial\varphi}{\partial\theta_{j'}}(z_{ip},\theta)e_{ijp}^{(1)}(\vec{z},\theta,\vec{k})$$

$$+\sum_{i=1}^{N}\sum_{p=1}^{n}\frac{\partial^2\varphi}{\partial\theta_j\partial\theta_{j'}}(z_{ip},\theta)e_{i0p}^{(2)}(\vec{z},\theta,\vec{k}) + e_{jj'}(\vec{z},\theta,\vec{k}).$$

Therefore from the assumption, we see that

$$V_{n0}(K,\vec{z},\theta) \subset U_0(\vec{z},\theta).$$

So we have the first assertion of Proposition 16 (1).

We remark that if

$$\frac{\partial\varphi}{\partial x}(z_{ip},\theta) > 0, \qquad \frac{\partial^2\varphi}{(\partial x)^2}(z_{ip},\theta) \neq 0,$$

for any $i = 1,\ldots,N$ and $p = 1,\ldots,n$, then

$$U_0(\vec{z},\theta) = V_{n0}(K,\vec{z},\theta).$$

Now note that

$$(M_i^{(n)I} K)(\vec{z}, \theta, \vec{k}) = e_0(\vec{z}, \theta, \vec{k}) + \sum_{p=1}^{n} z_{ip} \frac{\partial \varphi}{\partial x}(z_{ip}, \theta) e_{i0p}^{(1)}(\vec{z}, \theta, \vec{k}).$$

So we have $(\vec{M}^{(n)} K)_i(\vec{z}, \theta) - e_0(\vec{z}, \theta) \in V_{n0}(K, \vec{z}, \theta)$, $i \in I$ This implies that the second assertion of Proposition 16 (1).

Now let us prove the assertion (2) of Proposition 16. Suppose that K is an admissible family of copula functions. Then by Lemma 9, we see that there is a subset A of Θ such that the closure of A contains a non-void open subset of Θ and for any $\theta \in A$ there are $\xi_i \geqq 0$, $i = 1, \ldots, N$, such that $\sum_{i \in I} \xi_i > 0$ and $\sum_{i \in I} \xi_i (\vec{M}^{(n)} K)_i(\vec{z}, \theta) \in V_{n0}(K, \vec{z}, \theta)$. Then we see that $e_0(\vec{z}, \theta) \in V_{n0}(K, \vec{z}, \theta)$. This implies that $V_n(K, \vec{z}, \theta) \subset U_0(\vec{z}, \theta)$.

Then by the assertion (1), we see that $\dim V_n(K, \vec{z}, \theta) \leqq m_0$, $\theta \in A$. Let H_1 (resp. H_2) be the set of injective maps from $\{1, \ldots, m_0\}$ to C_n (resp. $\{1, \ldots, n\}^N$.) Now let

$$f(\theta) = \sum_{h_1 \in H_1} \sum_{h_2 \in H_2} \det(((\vec{M}^{(n)} K)_{h_1(r)})(\vec{z}, \theta, h_2(\ell)))_{r,\ell=1,\ldots,m_0})^2, \ \theta \in \Theta.$$

Then we see that $f(\theta) = 0$, $\theta \in A$. From the assumption, we see that $f : \Theta \to \mathbf{R}$ is real analytic. So we see that $f(\theta) = 0$, $\theta \in \bar{A} \cap \Theta$. Since $\bar{A} \cap \Theta$ contains a non-void open set and Θ is connected, we see that $f = 0$ on Θ. In particular, $f(\theta_0) = 0$. But this implies that $\dim V_n(K, \vec{z}, \theta_0) \leqq m_0 - 1$. This contradicts to the assumption. Therefore K is not admissible.

This completes the proof of Proposition 16.

9. Examples of dynamical default time copula models

Let $(\Omega, \mathcal{F}, P)$ be a complete probability space, $W(t) = (W^k(t))_{k=1,\ldots,d}$, $t \geqq 0$, be a d-dimensinal standard Wiener process. Let $N \geqq 2$, and $Z_1, \ldots, Z_N$ be a independent identically distributed random variables whose distributions are uniform distribution on $(0, 1)$. We assume that $\sigma\{Z_1, \ldots, Z_N\}$ and $\sigma\{W(t), t \geqq 0\}$ are independent. Let $M \geqq 1$. Let $\sigma_k : \mathbf{R}^M \to \mathbf{R}^M$, $k = 0, 1, \ldots, d$, be Lipschitz continuous functions and $h_i : \mathbf{R}^M \to (0, \infty)$, $i = 1, \ldots, N$, be continuous functions.

Let Y be the unique solution to the following stochastic differential equation on $\mathbf{R}^M$.

$$dY(t, y) = \sum_{k=1}^{d} \sigma_k(t, Y(t, y)) dW^k(t) + \sigma_0(t, Y(t, y)) dt,$$

$$Y(0, y) = y \in \mathbf{R}^M.$$

Let $y_0 \in \mathbf{R}^M$. We also assume that

$$P(\int_0^\infty h_i(Y(t, y_0))dt = \infty) = 1, \qquad i = 1, \dots, N,$$

and the support of the distribution of $Y(t, y_0)$ under $e^{-t} \otimes P(d\omega)$ contains non-empty open set.

Now let us define random times $\tau_1, \dots, \tau_N$ by

$$\tau_i = \inf\{t > 0;\ \exp(-\int_0^t h_i(Y(s, y_0))ds) < Z_i\}, \qquad i = 1, \dots, N.$$

Then we see that

$$(\prod_{i \in I} 1_{\{\tau_i > t\}}) P(\tau_i > t_i,\ i \in I | \mathcal{F}_t)$$

$$= (\prod_{i \in I} 1_{\{\tau_i > t\}}) E[\exp(-\sum_{i \in I} \int_t^{t_i} h_i(Y(s, y_0)ds) | Y(t, y_0)]$$

for $t, t_1, \dots .t_N \geqq 0$ with $t < \min\{t_i;\ i \in I\}$ (c.f. [1],[2],[4]). Let

$$H(s_1, \dots, s_N, y) = E[\exp(-\sum_{i=1}^N \int_0^{s_i} h_i(Y(r, y)dr)]$$

$$s_1, \dots, s_N \geqq 0,\ y \in \mathbf{R}^M,$$

and

$$H_i(s, y) = E[\exp(-\int_0^s h_i(Y(r, y)dr)]. \qquad i = 1, \dots, N,\ s \geqq 0,\ y \in \mathbf{R}^M.$$

Then $H_i(\cdot, y) : [0, \infty) \to (0, 1]$, $i = 1, \dots, N$, is strictly decreasing surjective function. So the inverse functions $H_i^{-1}(\cdot, y) : (0, 1] \to [0, \infty)$, $i = 1, \dots, N$, exist. Let $K : [0, 1]^N \times \mathbf{R}^M \to [0, 1]$ be given by

$$K(x_1, \dots, x_N, y) = \begin{cases} H(H_1^{-1}(\cdot, y)(x_1), \dots, H_N^{-1}(\cdot, y), y), & \text{if } x_1, \dots, x_N \in (0, 1], \\ 0, & \text{if one of } x_1, \dots, x_N = 0. \end{cases}$$

Then we have

$$(\prod_{i=1}^N 1_{\{\tau_i > t\}}) P(\tau_i > t_i,\ i = 1, \dots, N | \mathcal{F}_t)$$

$$= (\prod_{i=1}^N 1_{\{\tau_i > t\}}) K(P(\tau_1 > t_1 | \mathcal{F}_t), \dots, P(\tau_N > t_N, Y(t, y_0))\ a.s.$$

for any $t \geqq 0$, and $t_1, \dots, t_N \in [t, \infty)$.

So we see that K is an admissible family of copula functions.

References

1. Bielecki, T. R., Jeanblanc, M., Rutkowski, M.: Modeling and valuation of credit risk. Stochastic methods in finance. In: Lecture Notes in Math., **1856**, pp. 27–126, Springer, Berlin (2004)
2. Bielecki, T. R., Rutkowski, M.: Credit Risk: Modelling, Valuation and Hedging (Springer Finance). Springer-Verlag, Berlin (2002)
3. Björk, T., Christensen, B.: Interest rate dynamics and consistent forward rate curves. Mathematical Finance **9**, 323–348 (1999)
4. Chesney, M., Jeanblanc, M., Yor, M.: Mathematical Methods for Financial Markets (Springer Finance). Springer-Verlag London, Ltd., London (2009)
5. Kusuoka, S.: A remark on default risk models. In: Kusuoka, S., Maruyama, M. (eds.) Advances in Mathematical Economics vol. **1**, pp. 69–82, Springer (1999)

Adv. Math. Econ. 16, 85–109 (2012)

Advances in
MATHEMATICAL
ECONOMICS

Profits and exploitation: a reappraisal*

Naoki Yoshihara[1] **and Roberto Veneziani**[2]

[1] The Institute of Economic Research, Hitotsubashi University, Naka 2-1, Kunitachi, Tokyo 186-0004, Japan
(e-mail: yosihara@ier.hit-u.ac.jp)

[2] Department of Economics, Queen Mary University of London, Mile End Road, London E1 4NS, United Kingdom
(e-mail: r.veneziani@qmul.ac.uk)

Received: January 17, 2011
Revised: September 2, 2011

JEL classification: D31, D46, B51, B24

Mathematics Subject Classification (2010): 91B50, 91B15, 91B66

Abstract. This paper provides a mathematical analysis of the Marxian theory of the exploitation of labour in general equilibrium models. The two main definitions of Marxian exploitation in the literature, proposed by Morishima (1974) and Roemer (1982), respectively, are analysed in the context of general convex economies. It is shown that, contrary to the received view, in general these definitions do not preserve the so-called Fundamental Marxian Theorem (FMT), which states that the exploitation of labour is synonymous with positive profits. A new definition of Marxian labor exploitation is proposed, which is shown to preserve the FMT in general convex economies, in equilibrium.

Key words: exploitation, profits, reproducible solutions

1. Introduction

The derivation of a general relation between the exploitation of labour and the existence of profits, and the definition of an appropriate notion of general equilibrium have historically been central (and partly related) issues in

* We would like to thank John Roemer and two anonymous referees for useful comments and suggestions on an earlier draft. The usual disclaimer applies.

mathematical Marxian economics. From a theoretical viewpoint, this is due to the fundamental relevance of the theory of exploitation, and of the notion of the reproducibility of an economic system in Marxian approaches. From a formal viewpoint, this is explained by the difficulty of proving some core propositions in exploitation theory which Marx himself, and the early Marxists, took for evident.

To be precise, a key tenet of Marxist theory is the so-called Fundamental Marxian Theorem (hereafter, FMT) which establishes a correspondence between a positive profit rate and the existence of exploitation. This theorem was originally proved by Okishio (1963), and later named as such by Morishima (1973), in the context of simple Leontief economies with homogeneous labour. Yet, outside of stylised, linear two-class economies, both the appropriate definition of exploitation and the validity of the core insights of exploitation theory, including the FMT, are not uncontroversial, and indeed a number of approaches have been proposed in an attempt to generalise the FMT (see Yoshihara, 2010).

This has generated a substantial literature. Whereas the FMT was successfully extended to Leontief economies with heterogeneous labour by Fujimori (1982), Krause (1982), and others, in a famous book Steedman (1977) proved that the FMT does not hold in more general von Neumann economies. One solution was proposed by Morishima (1974) based on a new definition of exploitation, and focusing on von Neumann's notion of Balanced Growth Equilibrium. Yet Roemer (1981) showed that if Morishima's (1974) definition is adopted, the FMT does not hold, in general, in economies with a convex cone technology, if a different notion of equilibrium is adopted, namely that of *reproducible solution* (Roemer, 1980), unless some restrictions on the production set are imposed. Later, Roemer (1982) proposed an alternative definition of exploitation, which according to him would generalise the FMT (and other key Marxian propositions) to convex cone economies at reproducible solutions.

Various authors have objected to Morishima's (1974) and Roemer's (1982) definitions of exploitation on theoretical and exegetical grounds and other approaches have been proposed (for recent debates, see Veneziani, 2004; Flaschel 2010). Yet they remain the most prominent definitions in the literature, and this is to a significant extent due precisely to the formal derivation of the FMT. The relevance of the FMT, in fact, is such that although it is proved as a result, its epistemological status is that of a postulate: the appropriate definition of exploitation is widely considered to be one which preserves the FMT.[1] There are various reasons why the FMT is considered a key tenet of exploitation theory, and of Marxian economics in general. In the standard approach, the FMT is often interpreted as the formalisation of Marx's claim that profits are uniquely determined by exploitation in class divided economies. According to Roemer (1980, 1981), instead, the FMT captures the productiveness of capitalist economies and it provides the necessary and sufficient conditions for the existence of nontrivial general equilibria. From a normative perspective, however, the FMT is relevant because it captures a key insight of the theory of exploitation as the *unequal exchange of labour*, according to which exploitative relations are characterised by a difference between the amount of labour that an individual provides and the amount of labour contained in some relevant bundle that she does (or can) purchase with her income. From this perspective, exploitation theory focuses on inequalities in the distribution of income and labour, and thus it captures some core *inequalities of well-being freedom* (see Rawls, 1971; Sen, 1985), which are characteristic of capitalist economies. According to the FMT, profits play a key role in the generation of exploitation as the unequal exchange of labour (and thus of inequalities in well-being freedom), since they represent the way in which capitalists appropriate social surplus and social labour.

In this paper, the FMT is analysed in the context of general economies with convex cone production technologies, both assuming a representative agent and allowing for heterogeneous preferences over consumption. Within this general framework, the classic definitions of exploitation proposed by Morishima (1974) and Roemer (1982) are reconsidered, and it is shown that, contrary to the received wisdom, neither of them preserves the FMT in general. If the equilibrium notion is that of a reproducible solution, then the FMT does not hold under Roemer's definition, even if workers are identical and consume a uniform subsistence bundle. Further, if one allows for heterogeneity in workers' preferences, then the FMT does not hold under Morishima's definition, either. A new definition of exploitation is proposed which focuses on workers' *income*, rather than on their *consumption bundle*. This definition seems preferable on theoretical grounds, because the exploitation status of an agent is identified based on the objective features of an economy (including data on production, income, labour supply, etc.), rather than on the subjective and idiosyncratic factors driving consumption choices. As argued by Yoshihara and Veneziani (2011), Marxian exploitation theory is inherently objectivist in nature, and two agents with the same endowments, labour supply, and income should have the same exploitation status, regardless of

[1] According to Roemer (1982), the Class-Exploitation Correspondence Principle plays a similar epistemological role. This is briefly discussed in the concluding section below.

their choice of a consumption bundle. Furthermore, it is shown that if this definition is adopted, the FMT holds in general convex economies both with a representative agent and if heterogeneous preferences are allowed for.

This paper is related to a small but growing recent literature on exploitation theory. A conceptually related definition, for example, has been proposed, and axiomatically derived, by Yoshihara (2010) and Yoshihara and Veneziani (2009, 2010). Moreover, although this paper focuses on static perfectly competitive economies, the mathematical Marxian exploitation theory has been recently extended to models of contested exchange (Yoshihara, 1998) and to dynamic economies with intertemporally optimising agents (Veneziani, 2007).

The rest of the paper is organised as follows. Section 2 outlines the basic model. Section 3 introduces the two main received definitions of Marxian exploitation and proves that the FMT does not hold in general under either of them. Section 4 presents a new definition of exploitation and shows that it preserves the FMT in general. Section 5 concludes.

2. The Basic Model

An economy consists of a set H of agents, or households, who trade m commodities. Let $\mathbb{R}$ be the set of real numbers, and let $\mathbb{R}_+$ (resp., $\mathbb{R}_{++}$) be the set of nonnegative (resp., strictly positive) real numbers. Production technology is freely available to all agents, who can operate any activity in the production set P, which has elements of the form $\alpha = \left(-\alpha_0, -\underline{\alpha}, \overline{\alpha}\right)$, where $\alpha_0 \in \mathbb{R}_+$ is the direct labour input; $\underline{\alpha} \in \mathbb{R}^m_+$ are the inputs of the produced goods; and $\overline{\alpha} \in \mathbb{R}^m_+$ are the outputs of the m goods. Thus, elements of P are vectors in $\mathbb{R}^{2m+1}$. The net output vector arising from α is denoted as $\widehat{\alpha} \equiv \overline{\alpha} - \underline{\alpha}$. Let the vector with all components equal to zero be denoted as $\mathbf{0}$. The following assumptions on P hold throughout the paper.[2]

Assumption 0 (A0). P is a closed convex cone with vertex $\mathbf{0}$.

Assumption 1 (A1). For all $\alpha \in P$, if $\overline{\alpha} \geq \mathbf{0}$ then $\alpha_0 > 0$.

Assumption 2 (A2). For all $c \in \mathbb{R}^m_+$, there is an $\alpha \in P$ such that $\widehat{\alpha} \geqq c$.

A1 implies that labour is indispensable to produce any non-negative and nonzero output vector. A2 states that any non-negative commodity vector is producible as net output. It is worth stressing that A0∼A2 represent a significant

[2] For all vectors $x, y \in \mathbb{R}^p$, $x \geqq y$ if and only if $x_i \geqq y_i$ $(i = 1, \ldots, p)$; $x \geq y$ if and only if $x \geqq y$ and $x \neq y$; $x > y$ if and only if $x_i > y_i$ $(i = 1, \ldots, p)$. Note that vectors are columns unless otherwise specified.

generalisation of the standard linear technologies analysed in the literature on the FMT, which represent special cases of the model considered in this paper. For example, a von Neumann technology is defined by a tuple (B, A, L) where B is an $m \times n$ output matrix; A is an $m \times n$ input matrix; and L is a $1 \times n$ vector of labour coefficients, and the production possibility set $P_{(A,B,L)}$ is given by

$$P_{(A,B,L)} \equiv \left\{\alpha \in \mathbb{R}^{2m+1} \mid \exists x \in \mathbb{R}^n_+ : \alpha \leqq (-Lx, -Ax, Bx)\right\}.$$

$P_{(A,B,L)}$ is a closed convex cone in $\mathbb{R}^{2m+1}$ with $\mathbf{0} \in P_{(A,B,L)}$ and it satisfies A1~A2.

Although different assumptions concerning agents' behaviour will be considered below, assumptions A0~A2 will be retained throughout the paper. For the sake of simplicity, in what follows, economies whose production set satisfies A0~A2 will be referred to as general convex economies.

Given P, the set of production activities feasible with $\alpha_0 = k$ units of labour can be defined as follows:

$$P(\alpha_0 = k) \equiv \left\{\left(-\alpha_0, -\underline{\alpha}, \overline{\alpha}\right) \in P \mid \alpha_0 = k\right\},$$

and the set of net output vectors feasible with k units of labour is:

$$\widehat{P}(\alpha_0 = k) \equiv \left\{\widehat{\alpha} \in \mathbb{R}^m \mid \exists \alpha' \in P(\alpha_0 = k) : \overline{\alpha}' - \underline{\alpha}' \geqq \widehat{\alpha}\right\}.$$

For any set $X \subseteq \mathbb{R} \times ... \times \mathbb{R}$, $\partial X \equiv \left\{x \in X \mid \nexists x' \in X \text{ s.t. } x' > x\right\}$ is the frontier of X, and $X^\circ \equiv X \backslash \partial X$ is the interior of X.

Given a market economy, a (row) vector $p \in \mathbb{R}^m_+$ describes the price of each of the m commodities in the economy. The nominal wage rate is normalised to one. For any agent $\nu \in H$, let $\omega^\nu \in \mathbb{R}^m_+$ denote her initial endowments. In the literature on the FMT, it is assumed that the set of agents H can be partitioned into two disjoint subsets, namely the working class, denoted as W, which comprises agents with no initial endowments; and the set N of capitalists, who own at least some productive assets. Formally, $W = \{\nu \in H \mid \omega^\nu = \mathbf{0}\}$ and $N = \{\nu \in H \mid \omega^\nu \geq \mathbf{0}\}$. Further, it is assumed that workers are endowed with one unit of (homogeneous) labour.

For a given price vector p and wage rate $w = 1$, capitalists are assumed to maximise profits subject to their wealth constraint. Formally, each $\nu \in N$ solves:[3]

[3] Thus, noting that inputs are traded at the beginning of the period and outputs at the end, the optimisation programme can be interpreted as incorporating an assumption of *stationary expectations* on prices (see Roemer 1980; 1981, Chapter 2).

$$\text{choose } \alpha^{\nu} \in P \text{ to maximise } p\overline{\alpha}^{\nu} - \left(p\underline{\alpha}^{\nu} + \alpha_0^{\nu}\right)$$
$$\text{s.t. } p\underline{\alpha}^{\nu} + \alpha_0^{\nu} \leqq p\omega^{\nu}.$$

The set of production processes that solve this problem is denoted by $\mathcal{A}^{\nu}(p, 1)$. In line with standard classical political economy, it is assumed that capitalists do not work and do not consume: they use their revenues to accumulate for production in the next period.

Following the standard literature on the FMT, workers are assumed to supply a fixed amount of labour, equal to their labour endowment, and to be abundant relative to social productive assets. This assumption reflects the Marxian view that involuntary unemployment is a structural feature of capitalist economies. Workers' consumption behaviour, instead, will be specified below, where two different models are considered, by assuming first that workers consume a fixed subsistence bundle, and then relaxing this assumption by endowing them with heterogeneous demand functions for commodities.

3. The classic approaches to the FMT

3.1. Two definitions of exploitation

In Marxian theory, exploitation is conceived of as the unequal exchange of labour between agents: considering a worker $\mu \in W$, exploitative relations are characterised by systematic differences between the labour contributed by μ to the economy and the labour 'received' by μ, which is given by the amount of labour contained, or embodied, in some relevant consumption bundle(s). Therefore, for any bundle $c \in \mathbb{R}^m_+$, it is necessary to define the labour value (or labour content) of c. Outside of standard Leontief economies, the definition of the labour content of c is not obvious, and various definitions have, in fact, been proposed. In this section, the two most prominent definitions in the literature are presented, namely the definitions proposed by Morishima (1974; see also Roemer, 1981), and Roemer (1982).

In Morishima's (1974) approach, the labour value of a bundle of goods is independent of the equilibrium the economy is in. For any $c \in \mathbb{R}^m_+$, let

$$\phi(c) \equiv \left\{\alpha \in P \mid \widehat{\alpha} \geqq c\right\},$$

denote the set of activities that produce at least c as net output. Then:

Definition 1. *(Morishima, 1974): The labour value of a bundle c, $l.v.(c)$, is*

$$l.v.(c) \equiv \min\left\{\alpha_0 | \alpha = \left(-\alpha_0, -\underline{\alpha}, \overline{\alpha}\right) \in \phi(c)\right\}.$$

Therefore, if a worker spends one unit of labour and buys the bundle $c \in \mathbb{R}^m_+ \setminus \{\mathbf{0}\}$, the rate of labour exploitation is defined as follows:

Definition 2. *(Morishima, 1974): The rate of labour exploitation at a consumption bundle $c \in \mathbb{R}^m_+ \setminus \{\mathbf{0}\}$ is*

$$e(c) \equiv \frac{1 - l.v.(c)}{l.v.(c)}.$$

It is easy to see that $\phi(c)$ is non-empty by A2, and that the set $\{\alpha_0 \mid \alpha \in \phi(c)\}$ is bounded below by 0, by the assumption $\mathbf{0} \in P$ and A1. Thus, $l.v.(c)$ is well-defined since P is closed. Moreover, by A1, $l.v.(c)$ is positive whenever $c \neq \mathbf{0}$, so that $e(c)$ is well-defined.

Morishima (1974) showed that, in the balanced growth equilibrium of a von Neumann linear economy with joint production, the warranted rate of profit is positive if and only if the rate of exploitation in Definition 2 is positive. This result holds even if the von Neumann technology contains inferior production processes. Roemer (1981) proved that if the more general equilibrium concept of reproducible solution (see Definition 5 below) is adopted, then the FMT holds under Definition 2 in economies with general convex technologies, provided inferior processes are ruled out. Roemer (1981) called the latter assumption *Independence of Production* (see Assumption 3 below).

These results would seem to settle the issue of the validity of the FMT, and to provide strong support to Definition 2. According to Romer (1982), however, Definition 2 does not preserve other key axioms of Marxian exploitation theory (such as the correspondence between class and exploitation status) and this led him to propose an alternative definition. Unlike in Morishima (1974), the definition of labour value in Roemer (1982) depends on the specific equilibrium the economy is in. Given $(p, 1) \in \mathbb{R}^{m+1}_+$ and $\alpha \in P$, let $\pi(p, 1; \alpha) \equiv \frac{p\widehat{\alpha} - \alpha_0}{p\underline{\alpha} + \alpha_0}$ be the corresponding profit rate and let $\overline{P}(p, 1) \equiv \arg\max \{\pi(p, 1; \alpha) \mid \alpha \in P\}$ be the set of profit–rate-maximising production activities. Given $(p, 1) \in \mathbb{R}^{m+1}_+$ and a consumption bundle $c \in \mathbb{R}^m_+$ let

$$\phi(c; p, 1) \equiv \left\{\alpha \in \overline{P}(p, 1) \mid \widehat{\alpha} \geqq c\right\},$$

denote the set of profit-rate-maximising activities which produce at least c as net output. The labour value of a bundle c is defined as follows:

Definition 3. *(Roemer, 1982): The labour value of commodity vector c at $(p, 1) \in \mathbb{R}^{m+1}_+$, $l.v.(c; p, 1)$, is given by*

$$l.v.(c; p, 1) \equiv \min \left\{\alpha_0 \mid \alpha = \left(-\alpha_0, -\underline{\alpha}, \overline{\alpha}\right) \in \phi(c; p, 1)\right\}.$$

The rate of labour exploitation is stated in the next definition.

Definition 4. *(Roemer, 1982): The rate of labour exploitation at a consumption bundle* $c \in \mathbb{R}_+^m \setminus \{\mathbf{0}\}$ *and a price vector* $(p, 1) \in \mathbb{R}_+^{m+1}$ *is*

$$e\,(c;\, p, 1) \equiv \frac{1 - l.v.\,(c;\, p, 1)}{l.v.\,(c;\, p, 1)}.$$

It is easy to verify that $l.v.\,(c;\, p, 1)$ is well-defined if $\phi\,(c;\, p, 1)$ is not empty, and it is positive whenever $c \neq \mathbf{0}$, so that $e\,(c;\, p, 1)$ is well-defined. Also, $l.v.\,(c;\, p, 1) \geqq l.v.\,(c)$ so that $e\,(c;\, p, 1) > 0$ implies $e\,(c) > 0$.[4]

Although Roemer (1982) does not provide an explicit formal analysis of the FMT using Definition 4, he argues that Definition 4 is superior to the alternatives (including Definition 2) also because the FMT continues to hold, provided a suitable restriction on the production set is imposed (Roemer, 1982, p.158). To this issue we turn next.

3.2. Roemer's (1982) definition reconsidered

In this subsection, the standard model for the FMT is analysed. In line with the literature, it is assumed that every agent in the working class consumes an exogenously given *subsistence vector* $b \in \mathbb{R}_+^m$, per unit of labour supplied. Thus, in this subsection a convex economy is given by technology P, agents $H = W \cup N$, subsistence bundle b, and productive endowments $\Omega = \left(\omega^1, \omega^2, ..., \omega^{|N|}\right)$, and is denoted as $E(P, H, b, \Omega)$. The definition of equilibrium for $E(P, H, b, \Omega)$ can then be provided:

Definition 5. *(Roemer, 1981, Definition 2.5, p.41): A reproducible solution (RS) for the economy* $E(P, H, b, \Omega)$ *is a pair* $\left((p, 1)\,, (\alpha^\nu)_{\nu \in N}\right)$*, where* $p \in \mathbb{R}_+^m$*, such that:*

(a) $\forall \nu \in N$, $\alpha^\nu \in \mathcal{A}^\nu\,(p, 1)$ *(profit maximisation);*
(b) $\widehat{\alpha} \geqq \alpha_0 b$ *(reproducibility),*
where $\widehat{\alpha} \equiv \sum_{\nu \in N} (\overline{\alpha}^\nu - \underline{\alpha}^\nu)$ *and* $\alpha_0 \equiv \sum_{\nu \in N} \alpha_0^\nu$*;*
(c) $pb = 1$ *(subsistence wage); and*
(d) $\underline{\alpha} + \alpha_0 b \leqq \omega$ *(social feasibility),*
where $\underline{\alpha} \equiv \sum_{\nu \in N} \underline{\alpha}^\nu$ *and* $\omega \equiv \sum_{\nu \in N} \omega^\nu$.

---

[4] Note that $\phi\,(c;\, p, 1)$ may be empty; for instance, if $\overline{P}\,(p, 1)$ consists of production activities whose corresponding net outputs are semi-positive vectors, then $\phi\,(c;\, p, 1)$ is empty whenever $c > \mathbf{0}$, so that $l.v.\,(c;\, p, 1)$ cannot be defined. This issue need not concern us here since the set $l.v.\,(c;\, p, 1)$ is never empty whenever equilibrium price vectors are considered, as in the analysis below.

Part (a) is standard and needs no further comment. Part (b) states that net output in every sector should at least be sufficient for employed workers' total consumption. This amounts to requiring that social endowments do not decrease, because (b) is equivalent to $\omega - \left(\underline{\alpha} + \alpha_0 b\right) + \overline{\alpha} \geqq \omega$, where the right hand side is the social stocks at the beginning of the period, and the left hand side is the stocks at the beginning of next period. Given that workers are abundant relative to productive assets, part (c) states that unemployment drives the equilibrium real wage rate down to the subsistence level. Finally, part (d) requires that intermediate inputs and workers' consumption can be anticipated from current stocks, since wages are assumed to be paid in advance.

The existence of a RS is guaranteed by the following proposition.

Proposition 1. *(Roemer 1980; 1981): Let $b \in \mathbb{R}^m_{++}$. Under A0~A2, a RS exists for the economy $E(P, H, b, \Omega)$.*

Next, the assumption of independence of production is stated (see Roemer, 1981, p.47), which rules out inferior production processes from P:

Assumption 3 (A3). (Independence of Production) If $\left(-\alpha_0, -\underline{\alpha}, \overline{\alpha}\right) \in P$ and $\mathbf{0} \leqq c \leq \widehat{\alpha}$, then there exists $\left(-\alpha_0', -\underline{\alpha}', \overline{\alpha}'\right) \in P$ such that $\widehat{\alpha}' \geqq c$ and $\alpha_0' < \alpha_0$.

Under A3, it can be shown that the FMT holds when Definition 2 is adopted:
Proposition 2. *(Roemer, 1981, Theorem 2.11): Let $b \in \mathbb{R}^m_{++}$. Under A0~A3, the following statements are equivalent:*

(i) $e(b) > 0$;
(ii) there exists a RS yielding positive total profits;
(iii) all RS's yield positive total profits.

A first important point to note is that if Definition 4 is adopted instead, it is not obvious at all, even under A3, that the FMT continues to hold, because $e(b) > 0$ does not necessarily imply $e(b; p) > 0$.[5] Indeed, Roemer (1982, Chapter 5, p.158, footnote 6) suggested that A3 be modified to hold on $\overline{P}(p, 1)$ in order to prove the FMT using Definition 4. Formally:

Assumption 3′ (A3′). Let $(p, 1)$ be the price vector at a RS. If $\left(-\alpha_0, -\underline{\alpha}, \overline{\alpha}\right) \in \overline{P}(p, 1)$ and $\mathbf{0} \leqq c \leq \widehat{\alpha}$, then there exists $\left(-\alpha_0', -\underline{\alpha}', \overline{\alpha}'\right) \in \overline{P}(p, 1)$ such that $\widehat{\alpha}' \geqq c$ and $\alpha_0' < \alpha_0$.

---

[5] Note that if $(p, 1)$ is a RS, $e(b; p)$ is well-defined regardless of the property of $\overline{P}(p, 1)$. This is because in any RS, it is always true that $\phi(b; p, 1)$ is non-empty by the fact that for the aggregate social production $\alpha = \sum_{\nu \in N} \alpha^\nu$ under the RS, $\alpha \in \overline{P}(p, 1)$ and $\widehat{\alpha} \geqq \alpha_0 b$ hold.

Unlike A3, however, A3′ seems rather uncompelling. A3 requires the elimination of inferior activities from possible production sets: this is a reasonable technological restriction on the feasible set, which implies no significant loss of generality. A3′ eliminates a very large class of production sets, including, for example, all those sets in which there are RS's with a unique profit-maximising path. One example is given in the proof of Theorem 1 below (see the economy described in Figure 1), but A3′ also rules out the entire class of production sets whose corresponding net output sets, $\widehat{P}(\alpha_0 = 1)$, are representable by strictly concave and continuously differentiable functions in which *any* RS has a unique profit-maximising path. But this seems a rather significant loss of generality and the rationale for this restriction is quite unclear. Moreover, it is not possible in general to check whether or not a given economy satisfies A3′ before characterising the set of RS's of the economy.

This implies that the robustness of the FMT using Definition 4 should be checked under A3, without appealing to the ad hoc version suggested by Roemer (1982). The following theorem, however, shows that even if A3 is assumed, the FMT does not hold in general under Definition 4.

Theorem 1. *Under A0∼A3, there exists an economy $E(P, H, b, \Omega)$ such that every RS $\left((p, 1), (\alpha^{\nu})_{\nu \in N}\right)$ yields positive total profits but $e(b; p, 1) = 0$.*

Proof. **1. (The economy)** Let $m = 2$, $W = \{\mu\}$, and $N = \{\nu\}$. Let $b = (1, 1)$ and let $\omega = \omega^{\nu} = (2, 1)$. Finally, define the following production points:

$$\alpha^1 = \left(-\alpha_0^1, -\underline{\alpha}^1, \overline{\alpha}^1\right) = (-1, (-2, -1), (2, 3));$$
$$\alpha^2 = \left(-\alpha_0^2, -\underline{\alpha}^2, \overline{\alpha}^2\right) = (-1, (-1, 0), (3, 1)); \text{ and}$$
$$\alpha^3 = \left(-\alpha_0^3, -\underline{\alpha}^3, \overline{\alpha}^3\right) = (-1, (-1, -1), (4, 1)).$$

Then, let P be a *closed, convex cone* subset of $\mathbb{R}^5$ such that

1) $\mathbf{0} \in P$; and
2) $co\left\{\alpha^1, \alpha^2, \alpha^3\right\} = P(\alpha_0 = 1)$, where coX is the convex hull of a set X.

P satisfies all the assumptions on the production set, including A3.
2. (The equilibria) We will now characterise the set of RSs. Note that:

$$\forall p \in \Delta^2 \backslash \{(1, 0)\},\ p\omega < p\left(\underline{\alpha}' + b\right)\ (\forall \underline{\alpha}' \in co\left\{\underline{\alpha}^1, \underline{\alpha}^2, \underline{\alpha}^3\right\} \backslash \left\{\underline{\alpha}^2\right\}), (*)$$

where Δ^2 is the two-dimensional simplex. Note also that, by Definition 5(b), if $((p, 1), \alpha)$ is a RS, then $\widehat{\alpha} \geqq \alpha_0 b$. Let $\alpha^{12} \equiv \frac{1}{2}\alpha^1 + \frac{1}{2}\alpha^2$: then

$co\{\alpha^{12}, \alpha^2\} = \{\alpha' \in \partial P(\alpha_0 = 1) \mid \widehat{\alpha}' \geqq b\}$. Thus, by the convex cone property of P, if $((p,1),\alpha)$ is a RS, there exist $t \in (0,1]$ and $\alpha' \in co\{\alpha^{12}, \alpha^2\}$ such that $\alpha = t\alpha'$.

Let $\Delta^2(\alpha^2) \equiv \{p \in \Delta^2 \mid \frac{1}{3} \leqq p_1 \leqq \frac{1}{2}, \frac{2}{3} \geqq p_2 \geqq \frac{1}{2}\}$. If $p \notin \Delta^2(\alpha^2)$, it is immediate to check that there is no $\alpha' \in co\{\alpha^{12}, \alpha^2\}$ such that for some appropriate $t \in (0,1]$, $t\alpha'$ constitutes a profit maximiser at that price. Hence $p \notin \Delta^2(\alpha^2)$ cannot hold at a RS. Therefore consider $p \in \Delta^2(\alpha^2)$.

If $p = \left(\frac{1}{3}, \frac{2}{3}\right)$, then $p\widehat{\alpha}^1 - \alpha_0^1 = p\widehat{\alpha}^2 - \alpha_0^2 = p\widehat{\alpha}' - \alpha_0' > p\widehat{\alpha}'' - \alpha_0''$ for any $\alpha' \in co\{\alpha^1, \alpha^2\}$ and any $\alpha'' \in co\{\alpha^2, \alpha^3\} \setminus \{\alpha^2\}$. However, because of the property (*), the capital constraint in the profit maximisation problem implies that α^2 is the unique profit maximiser at the price $p = \left(\frac{1}{3}, \frac{2}{3}\right)$.

If $p = \left(\frac{1}{2}, \frac{1}{2}\right)$, then $p\widehat{\alpha}^2 - \alpha_0^2 = p\widehat{\alpha}'' - \alpha_0'' > p\widehat{\alpha}' - \alpha_0'$ for any $\alpha' \in co\{\alpha^1, \alpha^2\} \setminus \{\alpha^2\}$ and any $\alpha'' \in co\{\alpha^2, \alpha^3\}$. Thus, by the same reasons as in the above paragraph, α^2 is the unique profit maximiser at the price $p = \left(\frac{1}{2}, \frac{1}{2}\right)$ also.

If p is such that $\frac{1}{3} < p_1 < \frac{1}{2}, \frac{2}{3} > p_2 > \frac{1}{2}$, then α^2 is the unique profit maximiser at that price.

In sum, at any RS, $\alpha = \alpha^2$ must hold. Indeed, for any $p^* \in \Delta^2(\alpha^2)$, it is immediate to check that $((p^*, 1), \alpha^2)$ constitutes a RS. Moreover, in this case, $\pi(p^*, 1; \alpha^2) > 0$.

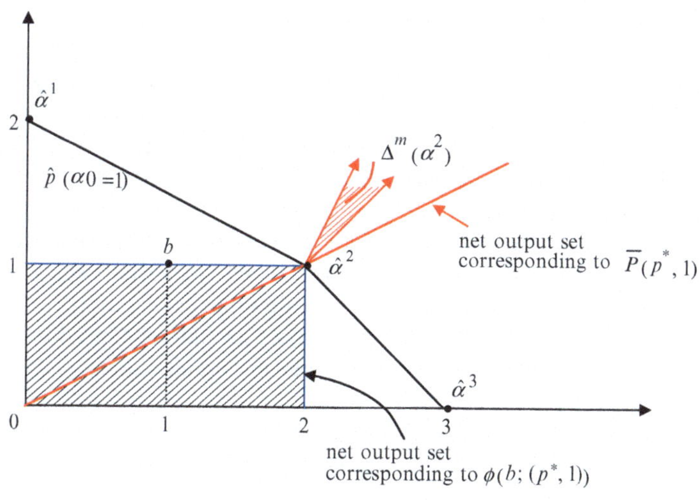

3. (FMT) We can now check that at any RS $((p^*, 1), \alpha^2)$, where $p^* \in \Delta^2(\alpha^2)$, $e(b; p^*, 1) = 0$ holds, whereas profits are positive. This is because $\overline{P}(p^*, 1) = \{t\alpha^2 \in P \mid t \in \mathbb{R}_{++}\}$ and $\phi(\alpha_0^2 b; p^*, 1) = \{\alpha^2\}$, which implies $l.v.(\alpha_0^2 b; p^*, 1) = \alpha_0^2$, so that $l.v.(b; p^*, 1) = 1$. ■

Remark. Note that in the above proof of Theorem 1, though the constructed economy has $b \in \mathbb{R}^m_{++}$, this proof can be applied even to the case of $b = (0, 1)$, so that the result does not depend on the assumption of $b \in \mathbb{R}^m_{++}$.

Theorem 1 proves that the FMT does not hold, in general, if Definition 4 is adopted, even under A3. It is actually worth noting that Theorem 1 does not only prove that there are economies in which at *some* equilibrium allocation the FMT does not hold. More strongly, and more significantly from a theoretical viewpoint, Theorem 1 shows that there are economies in which the FMT *never* holds in equilibrium if Roemer's (1982) definition is adopted. The intuition is the following: in the economy considered, at any RS $((p, 1), \alpha^2)$, the activity α^2 is the unique profit maximiser at p, and the corresponding net output $\widehat{\alpha}^2$ does not strictly dominate the subsistence bundle b, as described in Figure 1. In this case, the minimum amount of direct labour necessary to produce at least b (among profit-maximising production activities) is equal to the amount of direct labour expended at the RS.

To be sure, the economy considered in the proof of Theorem 1 does not satisfy A3′. But, as already noted, the theoretical relevance of A3′ is dubious and indeed one may wonder whether the reason of the failure of the FMT may lie elsewhere. As argued in the Introduction, the specific notion of equilibrium adopted is quite important in the analysis of the FMT: for example, if A3 is dropped, Morishima's (1974) definition allows one to derive the FMT if one focuses on balanced growth equilibria, but not if RSs are considered. Then, one may argue that the definition of exploitation is correct, but the notion of equilibrium adopted, namely the RS, is inappropriate.

Arguably, Definition 5 captures various key aspects of the Marxian notion of reproducibility of an economy, including its emphasis on social feasibility (condition (d)). The latter aspect, however, represents an important departure from standard Walrasian notions of equilibrium and therefore it may be worth considering whether the FMT holds under Definition 4 by adopting an equilibrium notion without the explicit capital constraint. Formally, let:[6]

Definition 5*. *A reproducible solution* (RS*) for the economy $E(P, H, b, \Omega)$ is a pair $((p, 1), (\alpha^\nu)_{\nu \in N})$, where $p \in \mathbb{R}^m_+$, such that conditions (b) and (c) of Definition 5 hold, and:*

[6] The equilibrium notion in Definition 5* is based upon Roemer (1981, pp.65-67).

(a)* $\forall \nu \in N$, $\alpha^\nu \in \mathcal{A}^{*\nu}(p,1)$ *(profit maximisation),*
where $\mathcal{A}^{*\nu}(p,1) \equiv \arg\max\left\{p\left(\widehat{\alpha}' - \alpha'_0 b\right) \mid \alpha' \in P \text{ and } \alpha'_0 \leqq W^\nu\right\}$
and W^ν *denotes* ν*'s financial endowment;*

(d)* $\alpha_0 \leqq \overline{L}$ *(social feasibility of labour demand),*
where $\alpha_0 \equiv \sum_{\nu \in N} \alpha_0^\nu$*, and* $\overline{L} \equiv \sum_{\nu \in N} W^\nu$.

The next Theorem provides partial support to the idea that the definition of equilibrium is relevant for establishing a correspondence between profits and exploitation.

Theorem 2. *Let* $b \in \mathbb{R}^m_{++}$. *Under A0~A3, for any economy* $E(P, H, b, \Omega)$*, the following statements hold:*

*(i) there exists a RS** $\left((p^*,1), (\alpha^\nu)_{\nu \in N}\right)$ *such that* $[p^*\left(\widehat{\alpha}^* - \alpha_0^* b\right) > 0 \Rightarrow e(b; p^*, 1) > 0]$;
*(ii) for any RS** $\left((p,1), (\alpha^\nu)_{\nu \in N}\right)$, $[e(b; p, 1) > 0 \Rightarrow p(\widehat{\alpha} - \alpha_0 b) > 0]$.

Proof. 1. By Theorem 2.17 in Roemer (1981; Chapter 2; Appendix 2), we know that a RS*, $\left((p,1), (\alpha^\nu)_{\nu \in N}\right)$, exists.

2. At any RS*, $\left((p,1), (\alpha^\nu)_{\nu \in N}\right)$, $\alpha_0 = \overline{L}$ by profit maximisation, without loss of generality, and p is an efficiency price which supports $\widehat{\alpha} \in \widehat{P}\left(\alpha_0 = \overline{L}\right)$ as an efficient production point. In other words, $p\widehat{\alpha} \geqq p\widehat{\alpha}'$ holds for any $\widehat{\alpha}' \in \widehat{P}\left(\alpha_0 = \overline{L}\right)$. Note that since $b \in \mathbb{R}^m_{++}$ and $\widehat{\alpha} \geqq \alpha_0 b$, then $\widehat{\alpha} \in \mathbb{R}^m_{++}$. Hence, by A3, $p \in \mathbb{R}^m_{++}$ holds and therefore at a RS*

$$p(\widehat{\alpha} - \alpha_0 b) > 0 \Leftrightarrow \widehat{\alpha} - \alpha_0 b \geq \mathbf{0}.$$

3. Show (ii). Take any RS*, $\left((p,1), (\alpha^\nu)_{\nu \in N}\right)$, and suppose $p(\widehat{\alpha} - \alpha_0 b) = 0$. Then, since $\widehat{\alpha} - \alpha_0 b = \mathbf{0}$ and $\widehat{\alpha} \in \partial\widehat{P}\left(\alpha_0 = \overline{L}\right)$, it follows that $l.v.(\alpha_0 b) = \overline{L}$. Hence, noting that $l.v.(\alpha_0 b; p, 1) \geqq l.v.(\alpha_0 b)$, it must be $l.v.(\alpha_0 b; p) \geqq \overline{L}$, so that $e(b; p, 1) \leqq 0$.

4. Show (i). Let $b \in \partial\widehat{P}(\alpha_0 = 1)$. Then, it follows from A3 that, for any $\alpha \in P$ with $\widehat{\alpha} - \alpha_0 b \geqq \mathbf{0}$, $\widehat{\alpha} - \alpha_0 b = \mathbf{0}$ holds. Therefore, for any RS*, $\left((p,1), (\alpha^\nu)_{\nu \in N}\right)$, $p(\widehat{\alpha} - \alpha_0 b) = 0$.

   Let $b \in \widehat{P}^\circ(\alpha_0 = 1)$. Then, there exists $\alpha^* \in \partial P$ such that $\widehat{\alpha}^* - \alpha_0^* b > \mathbf{0}$. Let $p^* \in \mathbb{R}^m_{++}$ be a price vector which supports $\widehat{\alpha}^* \in \partial\widehat{P}\left(\alpha_0 = \overline{L}\right)$ as an efficient production activity. Then, it is possible to construct a RS* $\left((p^*,1), (\alpha^{*\nu})_{\nu \in N}\right)$ such that $\sum_{\nu \in N} \alpha^{*\nu} = \alpha^*$. This is because it is easy to find a division $(\alpha^{*\nu})_{\nu \in N}$ of α^* such that $\alpha^{*\nu} \in \mathcal{A}^{*\nu}(p^*, 1)$ and $\sum_{\nu \in N} \alpha_0^{*\nu} = \overline{L}$, given that initial wealth for capitalists is only to finance the purchase of labour power. At this RS*, since $\widehat{\alpha}^* - \alpha_0^* b > \mathbf{0}$, $p^*\left(\widehat{\alpha}^* - \alpha_0^* b\right) > 0$ holds. Moreover, by A3, there exists $\alpha^{**} \in P$ such that $\widehat{\alpha}^{**} - \alpha_0^* b \geqq \mathbf{0}$ and $\alpha_0^{**} < \alpha_0^*$. In particular, since $\widehat{\alpha}^* - \alpha_0^* b > \mathbf{0}$, we can choose $\alpha^{**} = t\alpha^*$ with $t \in (0,1)$ by the cone property of P. In this case, $\alpha^{**} \in \overline{P}(p^*, 1)$ so that $l.v.\left(\alpha_0^* b; p^*, 1\right) < \alpha_0^*$. This implies $e(b; p^*, 1) > 0$, as desired. ■

Theorem 2 proves that for *every* convex cone economy, there always exists *one* equilibrium allocation (as defined in Definition 5*) such that the correspondence between profits and exploitation holds. This result would seem to establish the robustness of the FMT under Definition 4, without any ad hoc restrictions on A3, provided the appropriate equilibrium concept is adopted. This conclusion is unwarranted, though, as Theorem 2 proves a *weak* FMT. In fact, for every economy, the weakening of the capitalists' wealth constraints enlarges the set of equilibria and allows one to find one RS* such that the desired relation between profits and exploitation holds. Yet this is not true in general for *every* equilibrium allocation, even if Definition 5* is adopted, since equilibria similar to the one described in the proof of Theorem 1 are not ruled out.

Indeed, and this is the second limit of Theorem 2, it is possible that even if workers supply a constant amount of labour and receive the same bundle of wage goods, they may be exploited or not depending on the equilibrium aggregate production point. For instance, given $\alpha_0 > 0$ and $\alpha_0 b > \mathbf{0}$, if $\widehat{\alpha} - \alpha_0 b > \mathbf{0}$ at a RS*, then they are exploited, while if $\widehat{\alpha}' - \alpha_0 b \geq \mathbf{0}$ at another RS*, then economic relations may be nonexploitative. The latter situation may occur, for instance, if the boundary $\partial \widehat{P}(\alpha_0 = 1)$ is representable by a strictly concave and continuously differentiable function, because $\overline{P}\left(p', 1\right)$ corresponds to the set of vectors such that $\left(t\widehat{\alpha}', t\alpha_0\right)$ where $t \in \mathbb{R}_{++}$. This ambiguity seems rather dubious, at least if attention is restricted to equilibrium allocations in which capitalists maximise profits, since the exploitation status of workers should depend only upon the *objective* features of the labour contract, such as α_0 and $\alpha_0 b$.[7] Instead, the previous example suggests that the actual choice of the aggregate (equilibrium) net output vector may influence the workers' exploitation status even if their labour conditions are unchanged.

3.3. Morishima's (1974) definition and workers' heterogeneous consumption demands

The previous analysis suggests that the definition of exploitation proposed by Roemer (1982) is not superior to Morishima's (1974), at least as far as the

[7] If disequilibrium allocations are also considered, then capitalists' choices - which may turn out to be suboptimal ex post - may also be relevant for the determination of exploitation status. For a thorough discussion and an axiomatic defence of objectivism in the Marxian theory of exploitation, see Yoshihara and Veneziani (2011).

FMT is concerned. Indeed, it fares strictly worse: if A3 is imposed, the FMT holds in general at any RS of any convex economy $E(P, H, b, \Omega)$ under Definition 2, but not under Definition 4. The next question, then, concerns whether this result can be extended to more general economies. Although the technologies allowed for are very general (and standard in microeconomic theory), the assumptions on workers' behaviour seem restrictive, as they rule out both workers' choice of consumption bundles and heterogeneity in preferences. And both are among the important features of advanced economies that make the issue of exploitation a contentious one today.

In this subsection, the validity of the FMT is analysed in economies with heterogeneous workers, who have potentially different consumption demands. In standard Leontief economies, the assumption of heterogeneity of preferences has no implications for the validity of the FMT: a positive rate of profit prevails at a RS if and only if the average rate of exploitation of all workers is positive, which in turn holds if and only if each and every worker is exploited (see Roemer, 1981). The following analysis will show, however, that this result no longer holds once general convex economies with heterogeneous consumption demands are considered.

Let T be the nonempty, finite set of types of workers with heterogeneous consumption demands and let the generic element of T be denoted as τ. Let $F(\tau) \in [0, 1]$ be the fraction of workers of type τ. By definition, $\sum_{\tau \in T} F(\tau) = 1$. Given $p \in \mathbb{R}^m_+$, the consumption demand of the τ-type worker per unit of income is denoted as $d^\tau(p) \in \mathbb{R}^m_+$. For all $\tau \in T$, the demand function $d^\tau(\cdot)$ is assumed to be derived from a continuous, strictly monotonic, strictly quasi-concave, and homothetic utility function, and $pd^\tau(p) = 1$ for any $p \in \mathbb{R}^m_+$ normalised to $\sum_{j=1}^m p_j = 1$.

Further, for all $\tau \in T$, α_0^τ is the amount of labour spent by τ-type workers. Therefore given $p \in \mathbb{R}^m_+$ and an aggregate production plan $\alpha \in P$, and noting that workers are abundant relative to productive assets, aggregate labour demanded α_0 is equal to aggregate labour expended: $\sum_{\tau \in T} \alpha_0^\tau = \alpha_0$. Then, the average consumption demand of employed workers is defined by:

$$d\left(p; \left(\alpha_0^\tau\right)_{\tau \in T}\right) \equiv \frac{\sum_{\tau \in T} \alpha_0^\tau d^\tau(p)}{\alpha_0}.$$

Note that $p \cdot d\left(p; \left(\alpha_0^\tau\right)_{\tau \in T}\right) = 1$ by definition.

An economy is now specified by a list $E_T = \left(P; N; \Omega; T; (F(\tau))_{\tau \in T}; (d^\tau(\cdot))_{\tau \in T}\right)$, and in what follows it will be denoted simply as E_T for the sake of notational simplicity. The equilibrium notion for the economies E_T with heterogeneous workers' demands is formalised as follows:

Definition 6. *A reproducible solution*^{T} *(*RS^T*) for the economy* E_T *is a pair* $\left((p,1), (\alpha^\nu)_{\nu \in N}, (\alpha_0^\tau)_{\tau \in T}\right)$, *where* $p \in \mathbb{R}_+^m$, *such that:*

(a) $\forall \nu \in N$, $\alpha^\nu \in \mathcal{A}^\nu(p,1)$ *(profit maximisation);*
(b) $\widehat{\alpha} \geqq \alpha_0 d\left(p; (\alpha_0^\tau)_{\tau \in T}\right)$ *(reproducibility),*
where $\widehat{\alpha} \equiv \sum_{\nu \in N}(\overline{\alpha}^\nu - \underline{\alpha}^\nu)$ *and* $\alpha_0 \equiv \sum_{\nu \in N} \alpha_0^\nu = \sum_{\tau \in T} \alpha_0^\tau$;
(c) $\forall \tau \in T$, $pd^\tau(p) = 1$; *and*
(d) $\underline{\alpha} + \alpha_0 d\left(p; (\alpha_0^\tau)_{\tau \in T}\right) \leqq \omega$ *(social feasibility),*

where $\underline{\alpha} \equiv \sum_{\nu \in N} \underline{\alpha}^\nu$ *and* $\omega \equiv \sum_{\nu \in N} \omega^\nu$.

Definition 6 is similar to Definition 5, except that the aggregate consumption demands of employed workers are endogenous and possibly heterogeneous.

The next result proves the existence of a RS^T for these general economies.

Proposition 3. *For all* $\tau \in T$, *let* $d^\tau(\cdot)$ *be a strictly monotone, continuous demand function whose range is within the consumption set* $C \subseteq \mathbb{R}_+^m$. *Under A0~A2, a RS*^{T} *exists for the economy* E_T.

Proof. Straightforward modification of the proof of Corollary 2.8 in Roemer (1981, p.44). ■

The next Theorem establishes a preliminary result concerning the correspondence between the existence of positive profits and the exploitation of the *average* worker, under Definition 2.

Theorem 3. *Under A0~A3, in the economy* E_T, *let* $d^\tau(\cdot)$ *be derived from continuous, strictly monotonic, strictly quasi-concave, and homethetic utility function defined on the consumption set* $C \subseteq \mathbb{R}_+^m$, *which can be either unbounded, or sufficiently large but bounded. Let* $\left((p,1), (\alpha^\nu)_{\nu \in N}, (\alpha_0^\tau)_{\tau \in T}\right)$ *be a* RS^T *for the economy* E_T. *Then, total profits are positive if and only if* $e\left(d\left(p; (\alpha_0^\tau)_{\tau \in T}\right)\right) > 0$.

Proof. **(⇒):** Let $\left((p,1), (\alpha^\nu)_{\nu \in N}, (\alpha_0^\tau)_{\tau \in T}\right)$ be a RS^T with positive total profits:

$$p\widehat{\alpha} - \alpha_0 = p \cdot \left(\widehat{\alpha} - \alpha_0 d\left(p; (\alpha_0^\tau)_{\tau \in T}\right)\right) > 0.$$

Since $p \in \mathbb{R}_+^m$ and $\widehat{\alpha} \geqq \alpha_0 d\left(p; (\alpha_0^\tau)_{\tau \in T}\right)$ by Definition 6(b), the last strict inequality implies $\widehat{\alpha} \geq \alpha_0 d\left(p; (\alpha_0^\tau)_{\tau \in T}\right)$. Thus, by A3, $l.v.\left(\alpha_0 d\left(p; (\alpha_0^\tau)_{\tau \in T}\right)\right) < \alpha_0$. By the convex cone property of P, $l.v.\left(d\left(p; (\alpha_0^\tau)_{\tau \in T}\right)\right) < 1$, which implies $e\left(d\left(p; (\alpha_0^\tau)_{\tau \in T}\right)\right) > 0$.

(⇐): Since there is no RS^T with negative total profits, let $\left((p,1), (\alpha^\nu)_{\nu \in N}, (\alpha_0^\tau)_{\tau \in T}\right)$ be a RS^T with zero aggregate profits. Thus,

$p \cdot (\widehat{\alpha} - \alpha_0 d(p; (\alpha_0^\tau)_{\tau \in T})) = 0$. By Definition 6(b), $\widehat{\alpha} \geqq \alpha_0 d(p; (\alpha_0^\tau)_{\tau \in T})$. If for some commodity j, $\widehat{\alpha}_j - \alpha_0 d_j(p; (\alpha_0^\tau)_{\tau \in T}) > 0$, then it follows that $p_j = 0$. However, since every worker has strictly monotone preferences, $p_j = 0$ implies $\widehat{\alpha}_j - \alpha_0 d_j(p; (\alpha_0^\tau)_{\tau \in T}) < 0$, a contradiction. Thus, $\widehat{\alpha} = \alpha_0 d(p; (\alpha_0^\tau)_{\tau \in T})$.

Suppose $l.v.(\alpha_0 d(p; (\alpha_0^\tau)_{\tau \in T})) < \alpha_0$. Then, $l.v.(\widehat{\alpha}) < \alpha_0$, which implies that there exists $\alpha' \in P$ such that $\widehat{\alpha}' \geqq \widehat{\alpha}$ and $\alpha_0' < \alpha_0$. Because of the cone property of P, there is $\alpha'' \in P$ such that $\widehat{\alpha}'' = \frac{\alpha_0}{\alpha_0'}\widehat{\alpha}'$ and $\alpha_0'' = \alpha_0$. If $\widehat{\alpha} > 0$, then $\widehat{\alpha}'' > \widehat{\alpha}$, so that $\frac{\widehat{\alpha}}{\alpha_0} \in \widehat{P}^\circ(\alpha_0 = 1)$ by $\frac{\widehat{\alpha}''}{\alpha_0} \in \partial \widehat{P}(\alpha_0 = 1)$. Let $\widehat{\alpha}_i' = \widehat{\alpha}_i = 0$ for some i. In this case, $\widehat{\alpha}'' > \widehat{\alpha}$ does not hold. However, by A2, there exists $c \in \widehat{P}(\alpha_0 = 1) \cap \mathbb{R}^m_{++}$ which is sufficiently close to $\frac{\widehat{\alpha}''}{\alpha_0}$. Then, $c > \frac{\widehat{\alpha}}{\alpha_0}$ holds, so that $\frac{\widehat{\alpha}}{\alpha_0} \in \widehat{P}^\circ(\alpha_0 = 1)$. However, the last equation implies $d(p; (\alpha_0^\tau)_{\tau \in T}) \in \widehat{P}^\circ(\alpha_0 = 1)$, since $\widehat{\alpha} = \alpha_0 d(p; (\alpha_0^\tau)_{\tau \in T})$, thus $p[c - d(p; (\alpha_0^\tau)_{\tau \in T})] > 0$ for some $c \in \widehat{P}(\alpha_0 = 1) \cap \mathbb{R}^m_{++}$ with $c > \frac{\widehat{\alpha}}{\alpha_0}$. This is a contradiction, since the RS^T $((p, 1), (\alpha^\nu)_{\nu \in N}, (\alpha_0^\tau)_{\tau \in T})$ has zero aggregate profits. Thus, $l.v.(\alpha_0 d(p; (\alpha_0^\tau)_{\tau \in T})) = \alpha_0$, so that $e(d(p; (\alpha_0^\tau)_{\tau \in T})) = 0$. ■

Theorem 3 derives a general relation between profits and exploitation at the average, or aggregate, level. By Theorem 3, it is possible to conclude that the exploitation of the *working class* is a necessary and sufficient condition for the existence of exploitation. Yet, it is unclear that this result provides all the necessary information concerning the FMT. In fact, although workers here may have heterogeneous demand functions for consumption goods, they are identical in terms of their labour endowments and labour skills, their preferences for leisure, and other labour conditions. Therefore one may argue that a robust extension of the FMT should provide definite answers concerning the exploitation status of each and every worker. Although this issue is usually ignored in the literature, this is only due to the simplifying assumption of a representative worker. Arguably, the classic (albeit often implicit) understanding of the FMT is that it concerns *all* members of the working class, and not just the average worker.

So, the theoretically relevant question is whether, under Definition 2, the FMT holds *for each and every worker* in general convex economies with heterogeneous consumption demands. The next result provides a necessary and sufficient condition for the FMT to hold in this sense.

Theorem 4. *Under A0∼A3, let* $((p, 1), (\alpha^\nu)_{\nu \in N}, (\alpha_0^\tau)_{\tau \in T})$ *be a* RS^T *for the economy* E_T*. Then, the following two statements are equivalent:*
(i) total profits are positive if and only if $e(d^\tau(p)) > 0$ *for any* $\tau \in T$*;*
(ii) total profits are positive if and only if $d^\tau(p) \in \widehat{P}^\circ(\alpha_0 = 1)$ *for any* $\tau \in T$.

Proof. It suffices to prove that $d^{\tau}(p) \in \widehat{P}^{\circ}(\alpha_0 = 1)$ holds for any $\tau \in T$ if and only if $e(d^{\tau}(p)) > 0$ for any $\tau \in T$. First, let $d^{\tau}(p) \in \widehat{P}^{\circ}(\alpha_0 = 1)$ hold for any $\tau \in T$. This implies for any $\tau \in T$, there exists $\widehat{\alpha} \in \partial \widehat{P}(\alpha_0 = 1)$ such that $\widehat{\alpha} > d^{\tau}(p)$. Then, since P is a convex cone satisfying A2, there exists $\alpha^* \in P$ with $\alpha_0^* < 1$, such that $\widehat{\alpha}^* \in \partial \widehat{P}(\alpha_0 = \alpha_0^*)$ and $\widehat{\alpha}^* \geq d^{\tau}(p)$. This also implies $e(d^{\tau}(p)) > 0$ for any $\tau \in T$.

Conversely, suppose that there exists $\tau' \in T$ such that $d^{\tau'}(p) \notin \widehat{P}^{\circ}(\alpha_0 = 1)$. Then, either $d^{\tau'}(p) \in \partial \widehat{P}(\alpha_0 = 1)$ or $d^{\tau'}(p) \notin P(\alpha_0 = 1)$. If $d^{\tau'}(p) \notin P(\alpha_0 = 1)$, then for any $\alpha' \in P(\alpha_0 = 1)$, $\widehat{\alpha}' \not\geqq d^{\tau'}(p)$ holds. If $d^{\tau'}(p) \in \partial \widehat{P}(\alpha_0 = 1)$, then it implies together with A3 that, for any $\alpha' \in P(\alpha_0 = 1)$, $\widehat{\alpha}' \not> d^{\tau'}(p)$. In sum, for any $\alpha' \in P(\alpha_0 = 1)$, either $\widehat{\alpha}' = d^{\tau'}(p)$ or $\widehat{\alpha}' \not> d^{\tau'}(p)$ holds for this $\tau' \in T$. This implies $e\left(d^{\tau'}(p)\right) \leqq 0$ for this $\tau' \in T$. ■

Theorem 4 states that, at any RS^T, the correspondence between the existence of profits and the exploitation of every worker is equivalent to the correspondence between the existence of profits and the existence of some surplus labour - in the sense that each worker's optimal consumption bundle could be produced with strictly less labour than is actually supplied by the worker. Theorem 4 is interesting because it provides a general characterisation result, but also because it allows us to derive the main conclusion on Morishima's (1974) classic definition of exploitation. The next result, in fact, proves that if the latter definition is adopted, it is possible that profits are positive but some types of workers are not exploited.[8]

Corollary 1. *Under A0~A3, there exists an economy E_T such that there exists a* RS^T, $\left((p, 1), (\alpha^{\nu})_{\nu \in N}, (\alpha_0^{\tau})_{\tau \in T}\right)$, *which yields positive total profits but* $e\left(d^{\tau^*}(p)\right) < 0$ *for some* $\tau^* \in T$.

Proof. **1. (The economy)** Let $m = 2$, $N = \{\nu\}$, and $\omega = \omega^{\nu} = (2.5, 0.75)$. Let $T = \{\tau, \tau'\}$ with $F(\tau) = 0.5 = F(\tau')$, and let the demand functions $d^{\tau}(\cdot)$ and $d^{\tau'}(\cdot)$ be derived from continuous, monotonic, quasi-concave, and homothetic utility functions. Furthermore let:

$$d^{\tau}(p) = (0.5, 1.25) \text{ and } d^{\tau'}(p) = (2.5, 0.25) \text{ if } p = \left(\frac{1}{3}, \frac{2}{3}\right).$$

[8] In this paper, a negative exploitation rate simply implies that (some types of) workers are not exploited. In a more general model, however, one might define *exploiters* as those agents whose labour supplied is smaller than the value of their labour power. Based on such definition, Corollary 1 could be interpreted as suggesting that some propertyless workers are actually exploiters.

Next, define the following four production points:

$$\alpha^1 = \left(-\alpha_0^1, -\underline{\alpha}^1, \overline{\alpha}^1\right) = (-1, (-1, 0), (2.5, 1));$$

$$\alpha^2 = \left(-\alpha_0^2, -\underline{\alpha}^2, \overline{\alpha}^2\right) = (-1, (0, -1), (2.5, 1.5));$$

$$\alpha^3 = \left(-\alpha_0^3, -\underline{\alpha}^3, \overline{\alpha}^3\right) = (-1, (-1.5, 0), (1.5, 1.01)); \text{ and}$$

$$\alpha^4 = \left(-\alpha_0^4, -\underline{\alpha}^4, \overline{\alpha}^4\right) = (-1, (0, -1.5), (2.6, 1.5)).$$

The production possibility set of this economy, P, is a *closed, convex cone* subset of $\mathbb{R}^5$ such that

1) $\mathbf{0} \in P$;
2) $P(\alpha_0 = 1) = co\left\{\alpha^1, \alpha^2, \alpha^3, \alpha^4\right\}$.

P satisfies all the assumptions on the production set, including A3.

2. (Equilibrium) Given the above economy, we show that $((p^*, 1), \alpha^*) = \left(\left(\left(\frac{1}{3}, \frac{2}{3}\right), 1\right), \alpha^1\right)$ with $\alpha_0^{\tau} = \alpha_0^{\tau'} = 0.5$ is a RS^T. First, note that p^* is an efficiency price for α^*. Thus, since $p^*\underline{\alpha}^* + \alpha_0^* = p^*\omega^{\nu} < p^*\underline{\alpha} + \alpha_0$ for any $\alpha \in P(\alpha_0 = 1) \setminus \left\{\alpha^1\right\}$, it follows that $\alpha^* \in \mathcal{A}^{\nu}(p^*, 1)$, and Definition 6(a) holds. Next, since $d^{\tau}(p^*) = (0.5, 1.25)$ and $d^{\tau'}(p^*) = (2.5, 0.25)$, Definition 6(c) holds, too. Moreover, since $\alpha_0^{\tau} = \alpha_0^{\tau'} = 0.5$ then $d\left(p^*; \left(\alpha_0^{\tau}, \alpha_0^{\tau'}\right)\right) = (1.5, 0.75)$, so that $\widehat{\alpha}^* = (1.5, 1) \geq (1.5, 0.75) = \alpha_0^* \cdot d\left(p^*; \left(\alpha_0^{\tau}, \alpha_0^{\tau'}\right)\right)$, so that Definition 6(b) holds. Finally, $\underline{\alpha}^* + \alpha_0^* \cdot d\left(p^*; \left(\alpha_0^{\tau}, \alpha_0^{\tau'}\right)\right) = (2.5, 0.75) = \omega$, so that Definition 6(d) holds.

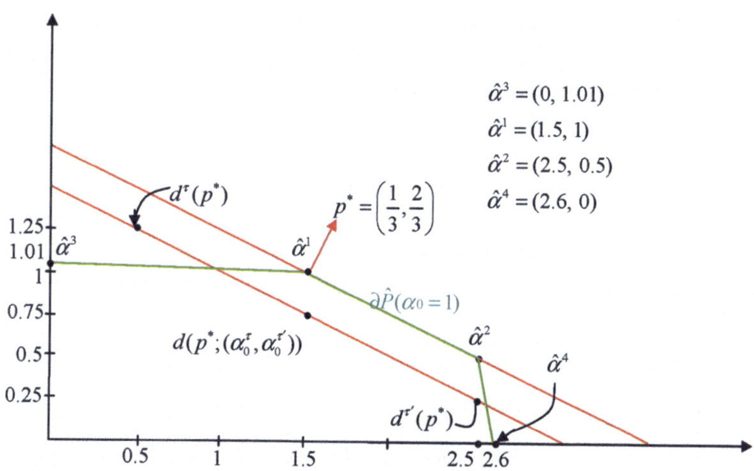

3. (FMT) This RS^T yields positive profits: $p^*\widehat{\alpha}^* - \alpha_0^* = \frac{1}{6} > 0$,[9] but employed workers of type τ are not exploited according to Definition 2, since $d^\tau(p^*) > \widehat{\alpha}^3$, and thus $d^\tau(p^*) \notin \widehat{P}^\circ(\alpha_0 = 1)$ which implies $e(d^\tau(p^*)) \leqq 0$, by Theorem 4. To prove that the latter inequality is strict, it suffices to note that $d^\tau(p^*) > \widehat{\alpha}^3$, with $l.v.(\widehat{\alpha}^3) = 1$, implies $l.v.(d^\tau(p^*)) > 1$. ■

Theorem 4 and Corollary 1 imply that outside of simple Leontief economies, *and even under A3*, Morishima's (1974) classic definition of exploitation does not preserve one of the key tenets of the Marxian theory of labour exploitation, namely the correspondence between profit making by capitalists and the exploitation of propertyless workers. The FMT does not hold in the sense that within the set of propertyless agents earning the same income and working the same amount of time, there may be some workers who are not exploited even if profits are positive. It is worth stressing the importance of the stronger interpretation of the FMT adopted: in the economy constructed in the proof of Corollary 1, workers spend the same amount of labour time with the same labour skills, earn the same wage rate, and face the same budget constraint. They are completely identical, except for the actual bundle consumed. In this situation, one would expect workers to have exactly the same exploitation status. Instead, Corollary 1 shows that if Definition 2 is adopted, some types of workers are exploited while others paradoxically emerge as exploiters in equilibrium, due to differences in their *subjective* consumption demands, even though all types of workers face the same *objective* labour conditions (identical labour supply, income, skills, and so on).

Finally, although Theorem 4 and Corollary 1 focus on the notion of RS^T formalised in Definition 6, similar results could be derived by focusing on Balanced Growth Equilibria as in Morishima (1974). The negative conclusions on the ability of Morishima's (1974) famous definition of exploitation to capture the core intuitions of Marxian exploitation theory do not depend on the specific equilibrium notion adopted.

4. A New Definition of Labour Exploitation

Given the epistemological status of the FMT in exploitation theory, the negative results derived in the previous section suggest that another definition

[9] Also, $e\left(d\left(p^*; \left(\alpha_0^\tau, \alpha_0^{\tau'}\right)\right)\right) > 0$, since $l.v.\left(d\left(p^*; \left(\alpha_0^\tau, \alpha_0^{\tau'}\right)\right)\right) < 1$. The last inequality follows from $l.v.(\widehat{\alpha}^*) = 1$, $\widehat{\alpha}^* \geq d\left(p^*; \left(\alpha_0^\tau, \alpha_0^{\tau'}\right)\right)$, and for any $\widehat{\alpha}' = \lambda\widehat{\alpha}^1 + (1-\lambda)\widehat{\alpha}^2 \in \widehat{P}(\alpha_0 = 1)$ with any $\lambda \in \left(0, \frac{1}{2}\right)$, $\widehat{\alpha}' > d\left(p^*; \left(\alpha_0^\tau, \alpha_0^{\tau'}\right)\right)$ holds.

must be found that preserves the core insights of Marxian theory. This is the task of this section. The key intuition of the following analysis is that, despite all of their differences, the definitions proposed by Morishima (1974) and Roemer (1982) suffer from the same conceptual and formal problem. For both approaches define the value of labour power - that is, the amount of labour 'received' by workers - based on the bundle of goods consumed by workers, thus making purely subjective and idiosyncratic factors central in exploitation theory. To be sure, in the analysis of the economy, the essential heterogeneity of workers should be taken into account, rather than assumed away as in the standard literature. Yet the definition of the value of labour power should depend on income, or purchasing power, rather than on the choice of a specific consumption bundle. Again, the notion of exploitation should be such that two agents who earn the same income by supplying the same amount of homogeneous labour are identified as having the same exploitation status, regardless of their consumption choices.

Formally, the labour value of a bundle of commodities is still given by Definition 3 above. However, the definition of the value of labour power focuses on the income received by workers and exploitation is measured by the difference between the (one unit of) labour supplied by every worker and the minimal amount of direct labour socially necessary to provide the agent with her *income* (per unit of labour). For any $\mu \in W$, let:

$$B(p, 1) \equiv \left\{ f^{\mu} \in \mathbb{R}^{m}_{+} \mid pf^{\mu} = 1 \right\}.$$

$B(p, 1)$ is the set of bundles that a worker can purchase, per unit of labour performed, using up all her income. Then:

Definition 7. *The rate of labour exploitation for worker* $\mu \in W$ *at* $(p, 1) \in \mathbb{R}^{m+1}_{+}$ *is*

$$e^{\mu}(p, 1) \equiv \frac{1 - \min_{f^{\mu} \in B(p,1)} l.v.(f^{\mu}; p, 1)}{\min_{f^{\mu} \in B(p,1)} l.v.(f^{\mu}; p, 1)}.$$

It is easy to verify that the value of labour power - $\min_{f^{\mu} \in B(p,1)} l.v.(f^{\mu}; p,1)$ - is well-defined, since for any $p \in \mathbb{R}^{m}_{+} \setminus \{\mathbf{0}\}$, there exists $f^{\mu} \in B(p, 1)$ such that $\phi(f^{\mu}; p, 1) \neq \varnothing$. Moreover, as already discussed, it has a positive value whenever $p \in \mathbb{R}^{m}_{+} \setminus \{\mathbf{0}\}$, so that $e^{\mu}(p, 1)$ is well-defined.

Thus, the value of labour power in Definition 7 focuses *not* on the worker's *consumption vector, but* rather on the *income* she earned at a RS. By Definition 7, worker μ is exploited at a given RS $\left((p, 1), (\alpha^{\nu})_{\nu \in N}\right)$ if and only if the minimum amount of labour socially necessary to provide μ with her (unit) labour income $w = 1$, is less than unity.

The next results show that a general, robust relation between positive profits and the exploitation of each and every worker can be derived under Definition 7. Theorem 5 focuses on the economy with homogeneous workers:

Theorem 5. *Under A0$\sim$A2, let $\left((p,1),(\alpha^{\nu})_{\nu\in N}\right)$ be a* RS *for an economy $E(P,H,b,\Omega)$. Then, total profits are positive if and only if $e^{\mu}(p,1)>0$ for every worker $\mu\in W$.*

Proof. (⇒): Let $\left((p,1),(\alpha^{\nu})_{\nu\in N}\right)$ be a RS with positive total profits: $p\widehat{\alpha}-\alpha_0 = p\cdot(\widehat{\alpha}-\alpha_0 b)>0$. Since $p\in\mathbb{R}^m_+$ and $\widehat{\alpha}\geqq\alpha_0 b$ by Definition 5(b), this implies $\widehat{\alpha}\geq\alpha_0 b$. Let $f\in\mathbb{R}^m_+$ be such that $pf=pb$ and $\alpha_0 f=t\widehat{\alpha}$ for some $0<t<1$. Note that such bundle exists, since $p\cdot(\widehat{\alpha}-\alpha_0 b)=p\cdot(\widehat{\alpha}-\alpha_0 f)>0$. Since $l.v.(\widehat{\alpha};p,1)\leqq\alpha_0$, it follows from $t\in(0,1)$ that $l.v.(t\widehat{\alpha};p,1)=l.v.(\alpha_0 f;p,1)<\alpha_0$. By linearity, $l.v.(f;p,1)<1$, which implies $\min_{\widehat{f}\in B(p,1)} l.v.\left(\widehat{f};p,1\right)<1$, so that $e^{\tau}(p,1)>0$ for every $\tau\in T$.
(⇐): Since there is no RS with negative total profits, let $\left((p,1),(\alpha^{\nu})_{\nu\in N}\right)$ be a RS such that $p\cdot(\widehat{\alpha}-\alpha_0 b)=0$. By Definition 5(b), $\widehat{\alpha}\geqq\alpha_0 b$. Therefore, if $p\in\mathbb{R}^m_{++}$, then $\widehat{\alpha}=\alpha_0 b$. Let $f\in\mathbb{R}^m_+$ be such that $pf=pb$ and $\alpha_0 f=t\widehat{\alpha}$ for some $0<t\leqq 1$. Then, $p\cdot(\widehat{\alpha}-\alpha_0 f)=0$ and $\alpha_0 f=t\widehat{\alpha}$ imply that $t=1$. Note that at this RS $\left((p,1),(\alpha^{\nu})_{\nu\in N}\right)$, any profit-rate-maximising production points $\alpha'\in\overline{P}(p,1)\cap\partial P(\alpha_0=1)$ has the property that $p\widehat{\alpha}'=1$ (by Definition 5(a) and the convex cone property of P). Thus, for any $\alpha'\in\overline{P}(p,1)\cap\partial P(\alpha_0=1)$, $p\widehat{\alpha}'=\frac{p\widehat{\alpha}}{\alpha_0}=pb$. This implies that for any $f\in\mathbb{R}^m_+$ such that $pf=pb$, $l.v.(f;p,1)\geqq 1$ holds. Hence, $\min_{\widehat{f}\in B(p,1)} l.v.\left(\widehat{f};p,1\right)=1$, so that $e^{\tau}(p,1)=0$ for every $\tau\in T$.

If $p\in\mathbb{R}^m_+$, it may be the case that $\widehat{\alpha}\geq\alpha_0 b$. However, as $p\cdot(\widehat{\alpha}-\alpha_0 b)=0$ and $\alpha^{\nu}\in\mathcal{A}^{\nu}(p,1)$ for all $\nu\in N$, $b\in\partial\widehat{P}(\alpha_0=1)$ holds. By the same argument as in the case with $p\in\mathbb{R}^m_{++}$, for any $f\in\mathbb{R}^m_+$ such that $pf=pb$, $l.v.(f;p,1)\geqq 1$ holds. Thus, $\min_{\widehat{f}\in B(p,1)} l.v.\left(\widehat{f};p,1\right)=1$, so that $e^{\tau}(p,1)=0$ for every $\tau\in T$. ■

It is worth noting that if Definition 7 is adopted, neither A3′ nor A3 are necessary to prove the FMT. This result is confirmed by the next Theorem, which establishes the robustness of the FMT also in economies with heterogeneous consumption demands.

Theorem 6. *Under A0$\sim$A2, let $\left((p,1),(\alpha^{\nu})_{\nu\in N},\left(\alpha_0^{\tau}\right)_{\tau\in T}\right)$ be a* RS^T *for E_T. Then, total profits are positive if and only if $e^{\tau}(p,1)>0$ for every $\tau\in T$.*

Proof. By Definition 6(c), at a RS^T, for any $\tau\in T$, $pd^{\tau}(p)=p\cdot d\left(p;\left(\alpha_0^{\tau}\right)_{\tau\in T}\right)=1$. Thus, for any $\tau,\tau'\in T$, $\min_{f^{\tau}\in B(p,1)} l.v.(f^{\tau};p,1)=\min_{f^{\tau'}\in B(p,1)} l.v.\left(f^{\tau'};p,1\right)$, so that $e^{\tau}(p,1)=e^{\tau'}(p,1)$ for any $\tau,\tau'\in T$. Let $\left((p,1),(\alpha^{\nu})_{\nu\in N}\right)$ be a RS^T with aggregate net output $\widehat{\alpha}$ and labour

expended α_0. Let $\widehat{\alpha}^0 \equiv \frac{\widehat{\alpha}}{\alpha_0}$. Then, as shown in the proof of Theorem 5, there exists some $t \in (0, 1]$ such that $pt\widehat{\alpha}^0 = 1$. Moreover, for any $\tau \in T$, $\min_{f^\tau \in B(p,1)} l.v.(f^\tau; p, 1) \leqq l.v.\left(t\widehat{\alpha}^0; p, 1\right)$.
($\Rightarrow$): Let $\left((p, 1), (\alpha^\nu)_{\nu \in N}, \left(\alpha_0^\tau\right)_{\tau \in T}\right)$ be a RS^T with positive total profits: $p\widehat{\alpha} - \alpha_0 = p \cdot \left(\widehat{\alpha} - \alpha_0 d\left(p; \left(\alpha_0^\tau\right)_{\tau \in T}\right)\right) > 0$. Thus, $p\widehat{\alpha}^0 > 1 = pt\widehat{\alpha}^0$, and so $1 > t$. The latter inequality implies $l.v.\left(t\widehat{\alpha}^0; p, 1\right) < 1$ by the cone property of P, since $\widehat{\alpha}^0 \in \partial \widehat{P}(\alpha_0 = 1)$ by profit maximisation. Thus, $\min_{\widehat{f} \in B(p,1)} l.v.\left(\widehat{f}; p, 1\right) < 1$, so that $e^\tau(p, 1) > 0$ for every $\tau \in T$.
($\Leftarrow$): Let $\left((p, 1), (\alpha^\nu)_{\nu \in N}\right)$ be a RS^T such that $p\widehat{\alpha} - \alpha_0 = p \cdot \left(\widehat{\alpha} - \alpha_0 d\left(p; \left(\alpha_0^\tau\right)_{\tau \in T}\right)\right) = 0$. Then, $p\widehat{\alpha}^0 = 1 = pt\widehat{\alpha}^0$, and so $t = 1$. Then, using the same argument as in the ($\Leftarrow$) part of the proof of Theorem 5, we can see that for any $f \in \mathbb{R}_+^m$ such that $pf = 1$, $l.v.(f; p, 1) \geqq 1$ and $l.v.\left(t\widehat{\alpha}^0; p, 1\right) = 1$ hold. Thus, $\min_{\widehat{f} \in B(p,1)} l.v.\left(\widehat{f}; p, 1\right) = 1$, so that $e^\tau(p, 1) = 0$ for every $\tau \in T$. ■

5. Conclusions

This paper analyses the mathematical Marxian theory of exploitation, focusing on the correspondence between positive profits and the existence of exploitation, and on the notion of reproducible solution. It is shown that, contrary to the received view, neither of the two main definitions of exploitation in the literature - proposed, respectively, by Morishima (1974) and Roemer (1982) - preserves the Fundamental Marxian Theorem in general convex economies. Given the central theoretical and epistemological role of the FMT in the Marxian theory of exploitation, this raises serious doubts on both approaches. The main shortcoming of the received definitions is their focus on the specific consumption bundle chosen by workers in the definition of the value of labour power (the amount of labour 'received' by workers in exchange for their own labour expenditure), which makes exploitation status dependent on purely subjective factors. From this perspective, the two main received definitions arguably fail to describe a fundamental feature of the Marxian theory, namely class struggle under historical materialism, since they make the exploitation concept dependent on workers' subjective views of the world: two workers with the same income, labour supply, and endowments may have different exploitation status based on their idiosyncratic choices as consumers.[10] An alternative definition is proposed, which defines the value of labour power based on workers' income. Under this new defini-

[10] We are grateful to an anonymous referee for pointing out the dubious relevance of subjective factors in Marxian theory.

tion a robust correspondence between positive profits and the exploitation of labour can be established in general convex economies with heterogeneous agents.

Two final remarks are worth making at this point about the robustness of our conclusions and the appeal of the proposed definition of exploitation. First, the economic models analysed in this paper are more general than those usually considered in the literature on Marxian exploitation. Yet they still contain a number of fairly strong simplifying assumptions, such as the neglect of consumption/leisure tradeoffs, workers' savings, capitalists' consumption, skills heterogeneity, and so on. One may legitimately wonder whether the FMT would still hold under the definition of exploitation proposed if the latter assumptions are violated. This topic is the object of ongoing work (Veneziani and Yoshihara, 2010), but preliminary results suggest that the main conclusions of this paper are robust. The theoretical and analytical emphasis on income earned, rather than actual consumption bundles, allows for a significant generalisation of standard insights.

Second, this paper focuses on the FMT as a key property of Marxian theory, but Roemer (1982) argued that an analogous epistemological role is played by the Class-Exploitation Correspondence Principle (CECP): any definition of exploitation should be such that agents in the upper classes emerge as exploiters and agents in the lower classes are exploited. Yoshihara (2010) has shown that neither Morishima's (1974) nor Roemer's (1982) definition preserves the CECP in general convex cone economies. In particular, an agent in the capitalist class may not be an exploiter if Morishima's (1974) definition is adopted; whereas an agent in the working class may not be exploited if Roemer's (1982) definition is adopted. Instead, the definition proposed in this paper does preserve the CECP in general (see Yoshihara, 2010), and thus it seems superior to standard approaches in this respect, too.

References

1. Flaschel, P.: Topics in Classical Micro- and Macroeconomics. Springer, New York (2010)
2. Fujimori, Y.: Modern Analysis of Value Theory. Springer-Verlag, Berlin (1982)
3. Krause, U.: Heterogeneous labour and the fundamental Marxian theorem. Review of Economic Studies **48**, 173–178 (1982)
4. Morishima, M.: Marx's Economics. Cambridge University Press, Cambridge (1973)

5. Morishima, M.: Marx in the light of modern economic theory. Econometrica **42**, 611–632 (1974)
6. Okishio, N.: A mathematical note on Marxian theorems. Weltwirtschaftsliches Archiv **91** (1963)
7. Rawls, J.: A Theory of Justice. Harvard University Press, Cambridge (1971)
8. Roemer, J. E.: A general equilibrium approach to Marxian economics. Econometrica **48**, 505–530 (1980)
9. Roemer, J. E.: Analytical Foundations of Marxian Economic Theory. Cambridge University Press, Cambridge (1981)
10. Roemer, J. E.: A General Theory of Exploitation and Class. Harvard University Press, Cambridge (1982)
11. Sen, A.K.: Commodities and Capabilities. North-Holland, Amsterdam (1985)
12. Steedman, I.: Marx after Sraffa. New Left Books, London (1977)
13. Veneziani, R.: The temporal single-system interpretation of Marx's economics: a critical evaluation. Metroeconomica **55**, 96–114 (2004)
14. Veneziani, R.: Exploitation and time. Journal of Economic Theory **132**, 189–207 (2007)
15. Veneziani, R., Yoshihara, N.: Exploitation and profits: a general axiomatic approach in convex economies with heterogeneous agents. In: IER Discussion Paper Series, No. A. 542, The Institute of Economic Research, Hitotsubashi University (2010)
16. Yoshihara, N.: Wealth, exploitation, and labor discipline in the contemporary capitalist economy. Metroeconomica **49**, 23–61 (1998)
17. Yoshihara, N.: Class and exploitation in general convex cone economies. Journal of Economic Behavior and Organization **75**, 281–296 (2010)
18. Yoshihara, N., Veneziani, R.: Exploitation as the unequal exchange of labour: an axiomatic approach. Working Paper No. 655, Queen Mary University of London (2009)
19. Yoshihara, N., Veneziani, R.: Strong subjectivism in the Marxian theory of exploitation: a critique. Metroeconomica **62**, 53–68 (2011)

Adv. Math. Econ. 16, 111–117 (2012)

Advances in
MATHEMATICAL
ECONOMICS

On the equivalence between the rejective core and the dividend equilibrium: a note*

Kentaro Miyazaki[1] **and Shin-Ichi Takekuma**[2]

[1] Graduate School of Economics, Hitotsubashi University, 2-1 Naka, Kunitachi, Tokyo 186-8601, Japan
(e-mail: ed081009@g.hit-u.ac.jp)

[2] Faculty of Economics, Hitotsubashi University, 2-1 Naka, Kunitachi, Tokyo 186-8601, Japan
(e-mail: takekuma.econ.hit-u@jcom.home.ne.jp)

Received: May 19, 2010
Revised: February 14, 2011

JEL classification: C71, D41, D51

Mathematics Subject Classification (2010): 28A20, 91A12, 91A13, 91B50

Abstract. Exchange economies in which preferences of some consumers are possibly satiated are considered. In a general model of an atomless exchange economy, the equivalence between the 'rejective' core and the set of dividend equilibrium allocations is proved by applying Liapunov's theorem in multi-dimensions.

Key words: rejective core, dividend equilibrium, core equivalence

1. Introduction

In general equilibrium theory, the non-satiation of consumers' preferences is an essential assumption and is assumed in proving the existence of the competitive equilibrium and the equivalence between the core and the competitive equilibrium. However, in recent analyses, it has been tried to relax the non-satiation assumption and to define several new notions of equilibrium for economies with satiation. Among them, the dividend equilibrium that was

* The authors are grateful to an anonymous referee for his very helpful comments which enabled us to simplify significantly the proof of the main theorem and to revise appropriately several expressions in this note.

originally defined by Aumann=Drèze [1] has been focused as a most natural and general notion.

The dividend equilibrium originates in the concept of 'coupons equilibrium' defined by Drèze=Müller [3] in the analysis of fixed price economies. In markets with price rigidities or in the more general context of markets with possibly satiated consumers, a Walrasian equilibrium may fail to exist. This led to a revision of the equilibrium concept and the dividend equilibrium was introduced. Aumann=Drèze [1] showed an equivalent relation between the dividend equilibrium and the Shapley value allocation.

Konovalov [6] showed that the classical equivalence result between a competitive equilibrium and the core, that was proved for atomless economies with non-satiation by Aumann [2], Hildenbrand [5], does not hold for a dividend equilibrium in the case of satiation, because the usual core is then too large. Konovalov introduced the notion of the "rejective core" and proved that in an atomless economy where consumers are equally distributed among a finite set of types, an allocation is a dividend equilibrium if and only if it belongs to the rejective core. The purpose of this note is to extend this result to the general case of an atomless economy where consumers may belong to a continuum of types. Our general proof follows the same line of argument as in Konovalov, but is based in an essential way on Liapunov's multi-dimensional theorem, as in Hildenbrand [4].

2. The model and the main theorem

We consider an exchange economy including L commodities and infinitely many consumers (continuum of consumers). The set of all consumers is denoted by a unit interval, $T = [0, 1]$. Each consumer $t \in T$ is characterized by a consumption set X_t, a preference relation $\succ_t$, and an initial endowment $w(t) \in X_t$. We assume that $X_t = \mathbb{R}^L_+$ for every t and that function $w : T \to \mathbb{R}^L_+$ is integrable [1].

An *assignment* is an integrable function $x : T \to \mathbb{R}^L_+$, in which $x(t)$ denotes the consumption bundle allocated to consumer $t \in T$. We call a fixed assignment, $w : T \to \mathbb{R}^L_+$, the *initial assignment*. An *allocation* is an assignment $x : T \to \mathbb{R}^L_+$ such that $\int_T x = \int_T w$ [2].

We assume the following conditions for each consumer $t \in T$.

(A1) $w(t)$ is in interior of $\mathbb{R}^L_+$.
(A2) $\succ_t$ is irreflexive.

[1] We denote by $\mathbb{R}^L$ an L-dimensional Euclidean space and by $\mathbb{R}^L_+$ its non-negative orthant.

[2] The integral of any assignment $x : T \to \mathbb{R}^L_+$ is denoted by $\int_T x$.

(A3) For each $\mathbf{x} \in \mathbb{R}_+^L$, set $\{\mathbf{y} \in \mathbb{R}_+^L | \mathbf{y} \succ_t \mathbf{x}\}$ is open in $\mathbb{R}_+^L$.

Furthermore, we need the following condition for mathematical treatments.

(A4) For any two assignments $x : T \to \mathbb{R}_+^L$ and $y : T \to \mathbb{R}_+^L$, set $\{t \in T | y(t) \succ_t x(t)\}$ is measurable.

Now, let us define the dividend equilibrium and the rejective core.

Definition 1. *An allocation $x : T \to \mathbb{R}_+^L$ is a dividend equilibrium allocation if there exist a price vector $\mathbf{p} \in \mathbb{R}^L \setminus \{\mathbf{0}\}$ and a measurable function $d : T \to \mathbb{R}_+$ such that for almost every $t \in T$,*

(i) $\mathbf{p} \cdot x(t) \leq \mathbf{p} \cdot w(t) + d(t)$,
(ii) $\mathbf{p} \cdot \mathbf{y} > \mathbf{p} \cdot w(t) + d(t)$ *for any* $\mathbf{y}$ *with* $\mathbf{y} \succ_t x(t)$.

Let us denote by $\mathscr{W}_d$ the set of dividend equilibrium allocations. In dividend equilibrium allocations, the surplus created by satiated consumers is distributed among non-satiated consumers as dividends which is denoted by function $d : T \to \mathbb{R}_+$. In Definition 1, if $d(t) = 0$ for almost all $t \in T$, a dividend equilibrium allocation $x : T \to \mathbb{R}_+^L$ is said to be a *competitive equilibrium allocation*. By $\mathscr{W}$, we denote the set of competitive equilibrium allocations. By definition, we have that $\mathscr{W} \subset \mathscr{W}_d$.

For a measurable subset S of $[0, 1]$, by $\lambda(S)$ we denote the Lebesgue measure of set S. When S is a measurable set of consumers whose Lebesgue measure is positive, we call it a *coalition*.

Definition 2. *A coalition S rejects an allocation $x : T \to \mathbb{R}_+^L$ if and only if there are a measurable partition $\{S_1, S_2\}$ of S and another allocation $y : T \to \mathbb{R}_+^L$, such that*

(i) $\displaystyle\int_S y = \int_{S_1} w + \int_{S_2} x,$
(ii) $y(t) \succ_t x(t)$ *for almost every* $t \in S$,
(iii) $w(t) \not\succ_t y(t)$ *for almost every* $t \in T \setminus S$.

Condition (i) allows each consumer t in coalition S to provide either his own initial endowment $w(t)$ or assignment $x(t)$ allotted to him in attaining a new allocation $y : T \to \mathbb{R}_+^L$. Condition (ii) means that the new allocation $y : T \to \mathbb{R}_+^L$ must be better than allocation $x : T \to \mathbb{R}_+^L$ for all consumers in coalition S. On the other hand, condition (iii) ensures that any individual consumer outside coalition S has no incentives to withdraw his initial endowment.

The *rejective core* is the set of all allocations that are not rejected by any coalition. Let us denote the rejective core by $\mathscr{C}_r$.

Konovalov [6] showed that an allocation is a dividend equilibrium if and only if it belongs to the rejective core in an atomless economy like in this paper, with the restriction that consumers are equally distributed among a finite set of types. We generalize here this equivalence result to the case of an atomless economy where consumers may belong to a continuum of types. Our general proof follows the same line of argument as in Konovalov, but relies in an essential way on Liapunov's multi-dimensional theorem, as in Hildenbrand [4].

Theorem 1. *Under assumptions (A1)-(A4), an allocation is a dividend equilibrium allocation if and only if it belongs to the rejective core, that is,* $\mathscr{W}_d = \mathscr{C}_r$.

3. Proof of the main theorem

The proof of the "only if" part of the theorem is identical to that of Konovalov [6] and is similar to the standard argument to prove that any Walras equilibrium allocation is a core allocation. Therefore, we skip the proof and only show that any allocation in the rejective core is a dividend equilibrium allocation.

Let $x : T \to \mathbb{R}^L_+$ be an allocation in the rejective core. Let S^* be the set of agents who are non-satiated in allocation $x : T \to \mathbb{R}^L_+$, i.e.,

$$S^* := \{t \in T \,|\, \mathbf{y} \succ_t x(t) \text{ for some } \mathbf{y} \in \mathbb{R}^L_+\}.$$

Then, it can be shown by (A3) that

$$S^* = \{t \in T \,|\, \mathbf{r} \succ_t x(t) \text{ for some } \mathbf{r} \in \mathbb{Q}^L_+\} = \bigcup_{\mathbf{r} \in \mathbb{Q}^L} \{t \in T \,|\, \mathbf{r} \succ_t x(t)\},$$

where $\mathbb{Q}^L$ is the set of all rational points in $\mathbb{R}^L$. Thus, by (A4), set S^* is measurable.

When S^* has measure 0, we have nothing to prove. In fact, we can easily see that x is a dividend equilibrium allocation associated with any price vector $\mathbf{p} \neq 0$ and a measurable function $d : T \to \mathbb{R}_+$ defined by $d(t) = \max\{0, \mathbf{p} \cdot x(t) - \mathbf{p} \cdot w(t)\}$. Thus we consider the case in which S^* has positive measure.

Now, for each $t \in T$, let us define the following sets:

$$G_w(t) := \{\mathbf{z} \in \mathbb{R}^L \,|\, w(t) + \mathbf{z} \succ_t x(t)\},$$
$$G_x(t) := \{\mathbf{z} \in \mathbb{R}^L \,|\, x(t) + \mathbf{z} \succ_t x(t)\}.$$

Then, by (A3), $G(t) = G_w(t) \cup G_x(t)$ is non-empty and open for all $t \in S^*$.

Following the technique of Hildenbrand [4], we define a subset of $\mathbb{R}^L$ by:

$$\Delta := \left\{ \int_E z \,\middle|\, z \in \mathscr{L}_G,\ E \text{ is a measurable subset of } S^* \text{ such that } \lambda(E)>0 \right\}.^3$$

Then, the set Δ is convex by Liapunov's theorem (Cor. 2 in Hildenbrand [4] p.451, due to Vind [7]). Also, we can easily show that Δ is non-empty (see Appendix, Claim 1).

We can prove the following lemma on the set Δ whose proof is very close to the argument by Konovalov [6].

Lemma. $\mathbf{0} \notin \Delta$.

Proof. Suppose that $\mathbf{0} \in \Delta$. Then, by definition of Δ, there is a measurable set $E^* \subset S^*$ with $\lambda(E^*) > 0$ and a measurable function z^* such that $z^*(t) \in G(t)$ a.e. $t \in E^*$ and $\int_{E^*} z^* = \mathbf{0}$. Here, define sets S_w and S_x by:

$$S_w := \{t \in E^* | z^*(t) \in G_w(t)\} \quad \text{and} \quad S_x := \{t \in E^* | z^*(t) \in G_x(t)\} \setminus S_w.$$

Then, it follows from (A4) that $\{S_w, S_x\}$ is a measurable partition of E^*.

Now again, by Liapunov's theorem, there are two measurable sets $S_1 \subset S_w$ and $S_2 \subset S_x$ respectively such that

$$\int_{S_1} (w, x, z^*) = \frac{1}{2} \int_{S_w} (w, x, z^*) \text{ and } \int_{S_2} (w, x, z^*) = \frac{1}{2} \int_{S_x} (w, x, z^*). \quad (1)$$

Similarly, by Liapunov's theorem, there is a measurable partition $\{V_1, V_2\}$ of $T \setminus E^*$ such that

$$\int_{V_1} (w, x) = \int_{V_2} (w, x) = \frac{1}{2} \int_{T \setminus E^*} (w, x). \quad (2)$$

Using these sets we define an assignment $y : T \to \mathbb{R}^L_+$ by

$$y(t) := \begin{cases} w(t) + z^*(t) & \text{when } t \in S_1, \\ x(t) + z^*(t) & \text{when } t \in S_2, \\ w(t) & \text{when } t \in V_2 \cup (S_x \setminus S_2), \\ x(t) & \text{when } t \in V_1 \cup (S_w \setminus S_1). \end{cases}$$

Let $S = S_1 \cup S_2$. Since $\lambda(S_w) > 0$ or $\lambda(S_x) > 0$, it follows that $\lambda(S_1) > 0$ or $\lambda(S_2) > 0$, i.e., $\lambda(S) > 0$. We will show that assignment $y : T \to \mathbb{R}^L_+$ can be a counter proposal to assignment $x : T \to \mathbb{R}^L_+$ by coalition S.

[3] $\mathscr{L}_G$ is the set of all integrable functions $z : T \to \mathbb{R}^L$ such that $z(t) \in G(t)$ a.e. in T.

By equations (1) and (2), we have

$$\begin{aligned}\int_T y &= \int_S z^* + \left(\int_{S_1} w + \int_{V_2} w + \int_{S_x \setminus S_2} w\right) + \left(\int_{S_2} x + \int_{V_1} x + \int_{S_w \setminus S_1} x\right)\\ &= \frac{1}{2}\left(\int_{S_w} w + \int_{T\setminus E^*} w + \int_{S_x} w\right) + \frac{1}{2}\left(\int_{S_x} x + \int_{T\setminus E^*} x + \int_{S_w} x\right)\\ &= \frac{1}{2}\left(\int_T w + \int_T x\right) = \int_T w,\end{aligned}$$

because $\int_S z^* = \frac{1}{2}\int_{E^*} z^* = \mathbf{0}$. This implies that $y : T \to \mathbb{R}_+^L$ is an allocation. Furthermore, we have $\int_S y = \int_S z^* + \int_{S_1} w + \int_{S_2} x = \int_{S_1} w + \int_{S_2} x$, which is condition (i) of Definition 2.

When $t \in S$, by construction of y, we can see that $y(t) \succ_t x(t)$ for a.e. $t \in S$, which is condition (ii) of Definition 2. When $t \in T \setminus S$, by construction of y, $y(t)$ equals $w(t)$ or $x(t)$. Consider the following two sets, $\{t \in T | w(t) \succ_t w(t)\}$ and $\{t \in T | w(t) \succ_t x(t)\}$. The first set has measure zero by irreflexivity of $\succ_t$. If the second set has positive measure, then the coalition consisting of consumers in the set rejects allocation x via the initial assignment w. Therefore, since x is an allocation in the rejective core, the second set cannot have positive measure. Thus, $w(t) \not\succ_t y(t)$ for a.e. $t \in T \setminus S$, which is condition (iii) of Definition 2.

Hence, we have shown that coalition S rejects an allocation x, which is a contradiction.

In what follows, we will show that allocation x is a dividend equilibrium allocation. By virtue of Lemma, we can apply Minkowski's separation theorem, and there is a vector $\mathbf{p} \in \mathbb{R}^n \setminus \{\mathbf{0}\}$ such that $\mathbf{p} \cdot \mathbf{z} \geq 0$ for any $\mathbf{z} \in \Delta$. This implies that for a.e. $t \in S^*$, $\mathbf{p} \cdot \mathbf{z} \geq 0$ for all $\mathbf{z} \in G(t)$ (see Appendix, Claim 2). Define a measurable function $d : T \to \mathbb{R}_+$ by $d(t) = \max\{0, \mathbf{p} \cdot x(t) - \mathbf{p} \cdot w(t)\}$. Then immediately we have $\mathbf{p} \cdot x(t) \leq \mathbf{p} \cdot w(t) + d(t)$ for each t, which is condition (i) of Definition 1. When $t \in S^*$, take any commodity vector $\mathbf{y} \in \mathbb{R}^L$ such that $\mathbf{y} \succ_t x(t)$. Then, $\mathbf{y} - w(t) \in G(t)$ and $\mathbf{y} - x(t) \in G(t)$. Therefore, $\mathbf{p} \cdot \mathbf{y} \geq \max\{\mathbf{p} \cdot w(t), \mathbf{p} \cdot x(t)\} = \mathbf{p} \cdot w(t) + d(t)$. In addition, under assumptions (A1) and (A3), we can easily show that $\mathbf{p} \cdot \mathbf{y} > \mathbf{p} \cdot w(t) + d(t)$. Since consumer t in $T \setminus S^*$ is satiated at $x(t)$, this implies condition (ii) of Definition 1.

This completes the proof of the main theorem.

4. Appendix

In this appendix, as a supplemental argument to the proof of the main theorem in Sect. 3, we will prove the following claims, which can be generally proved by using Measurable Choice Theorem (Hildenbrand [4], p. 452). However, under assumption (A3) and in our specific model where every consumer's consumption set is the same $\mathbb{R}^L_+$, we can easily prove them.

Claim 1. $\Delta \neq \emptyset$.

Claim 2. For a.e. $t \in S^*$, $\mathbf{p} \cdot \mathbf{z} \geq 0$ for all $\mathbf{z} \in G(t)$.

For each $\mathbf{z} \in \mathbb{R}^L$, define a set by $G^{-1}(\mathbf{z}) := \{t \in T | \mathbf{z} \in G(t)\}$. Then, by (A4), set $G^{-1}(\mathbf{z})$ is measurable. Let us define the following sets:

$$N := \left\{\mathbf{r} \in \mathbb{Q}^L | \lambda \left(G^{-1}(\mathbf{r})\right) = 0\right\} \quad \text{and} \quad U := S^* \setminus \bigcup_{\mathbf{r} \in N} G^{-1}(\mathbf{r}).$$

Since N is countable, $\lambda \left(\bigcup_{\mathbf{r} \in N} G^{-1}(\mathbf{r})\right) = 0$, and therefore $\lambda(U) = \lambda(S^*)$.

Proof of Claim 1. Let $t \in U$. Then, by (A3), there is an $\mathbf{r} \in \mathbb{Q}^L$ such that $\mathbf{r} \in G(t)$. By definition of U, $\mathbf{r} \notin N$, and therefore $\lambda \left(G^{-1}(\mathbf{r})\right) > 0$. If we put $E = G^{-1}(\mathbf{r})$, then $\lambda(E) > 0$ and $\mathbf{r} \in G(t)$ for all $t \in E$. Therefore, $\int_E \mathbf{r} \in \Delta$, which implies Claim 1.

Proof of Claim 2. Let $t \in U$ and $\mathbf{r} \in G(t) \cap \mathbb{Q}^L$. Then, by letting $E = G^{-1}(\mathbf{r})$, similarly we have $\int_E \mathbf{r} \in \Delta$. Thus, $\mathbf{p} \cdot \int_E \mathbf{r} \geq 0$, and therefore $\mathbf{p} \cdot \mathbf{r} \geq 0$. By (A3), this implies that $\mathbf{p} \cdot \mathbf{z} \geq 0$ for all $\mathbf{z} \in G(t)$. Thus, since $\lambda(U) = \lambda(S^*)$, Claim 2 follows.

References

1. Aumann, R.J., Drèze, J.H.: Values of markets with satiation or fixed prices. Econometrica **54**, 1271–1318 (1986)
2. Aumann, R.J.: Markets with a continuum of traders. Econometrica **32**, 39–50 (1964)
3. Drèze, J.H., Müller, H.: Optimality properties of rationing schemes. Journal of Economic Theory **23**, 131–149 (1980)
4. Hildenbrand, W.: The core of an economy with a measure space of economic agents. Review of Economic Studies **35**, 443–452 (1968)
5. Hildenbrand, W.: Core and Equilibria of a Large Economy, Princeton. Princeton University Press (1974)
6. Konovalov, A.: The core of an economy with satiation. Economic Theory **25**, 711–719 (2005)
7. Vind, K.: Edgeworth-allocations in an exchange economy with many traders. International Economic Review **5**, 165–177 (1964)

Adv. Math. Econ. 16, 119–131 (2012)

Advances in
MATHEMATICAL
ECONOMICS

Discrete stochastic calculus and its applications: an expository note*

Takahiko Fujita[1], **Naoyuki Ishimura**[2], **and Norihisa Kawai**[3]

[1] Information and Systems Engineering, Faculty of Science and Engineering, Chuo University, Kasuga, Tokyo 112-8551, Japan
(e-mail: rankstatistics@gmail.com)

[2] Graduate School of Economics, Hitotsubashi University, Kunitachi, Tokyo 186-8601, Japan
(e-mail: ishimura@econ.hit-u.ac.jp)

[3] Manulife Japan, Kokuryo, Chofu 182-8621, Japan
(e-mail: norihisa_kawai@manulife.com)

Received: October 28, 2011
Revised: September 5, 2011

JEL classification: C6, G1

Mathematics Subject Classification (2010): 60J65, 91G80

Abstract. We recast our recent studies on discrete stochastic processes relevant to a discrete analogue of the Itô formula. This analogous formula for discrete environment is introduced by one of the authors, and has a possibility of many applications in the discrete world. We consider the optimal portfolio problem and the pricing of exchange options. The results indicate certain direct connection between the discrete and the continuous processes through the Itô formula.

Key words: discrete analogue of the Itô formula, optimal portfolio problem, discrete Hamilton-Jacobi-Bellman equation, pricing of exchange options

* We are grateful to Professor Shigeo Kusuoka for precious suggestions, which helps improving the manuscript. The project of this research is motivated in part by the program "Financial Engineering Education for Graduate Course Students in Social Sciences (program leader: Professor Tsunemasa Shiba)," from the Ministry of Education, Science, Sports and Culture, Japan.

1. Introduction

This is an expository article concerning our recent studies on discrete stochastic calculus involving a discrete analogue of the famous Itô formula in continuous stochastic processes.

The celebrated Itô formula is known to provide a fundamental tool for the treatment of stochastic processes in continuous time. From the theory of stochastic analysis to the applications in finances, the formula designs a basic guideline for researches. One of the authors, on the other hand, recently formulated a discrete correspondence [3][5], which we call discrete analogue of the Itô formula. In terms of this analogous fourmula, the relation between the continuous and the discrete settings may be bridged rather easily. Moreover, wide variety of applications in the discrete world are expected.

In the present paper, we are concerned with such applications of the discrete analogue of the Itô formula. We here treat two examples: The optimal portfolio problem in discrete processes and the pricing of discrete Margrabe options. In the first example, the characterization of the optimality naturally leads to the discrete Hamilton-Jacobi-Bellman (d-HJB) equation. We exhibit several exact solutions. In the latter example, Margrabe options are also referred to as the exchange options, which give the holder of the option the right to exchange one for the other. We derive the pricing of such options under discrete random environment. Both examples show a kind of combinatorial strcture of the problem, and we believe that our method is suitable to computation.

The paper is organized as follows. In §2 we recall our discrete analogue of the Itô formula and present our underlying model. §3 is devoted to the study of optimal portfolio problem, whose presentation is similar to [7] but the examples are deepened. The pricing of Margrabe options is investigated in §4. We conclude with discussions in §5.

2. Discrete analogue of the Itô formula

We first recall the basic tool of our researches, namely, a discrete analogue of the Itô formula, which is formulated in the current form by Fujita and Kawanishi [5] (see also Fujita [3]).

Let $t = 0, 1, 2, \cdots$ denote discrete time series and let $\{B_t\}_{t=0,1,2,\cdots}$ with $B_0 = 0$ be the one-dimensional random walk:

$$B_t = \sum_{n=1}^{t} Z_n, \tag{1}$$

where $\{Z_n\}_{n=1,2,\cdots}$ are independent and identically distributed (i.i.d.) random variables. We here assume, for simplicity,

$$P(Z_n = 1) = P(Z_n = -1) = \frac{1}{2}, \qquad n = 1, 2, \cdots. \tag{2}$$

Precisely stated, we confine ourselves to treating the symmetric random walk. Generalizations are possible (see for instance [3][7]).

The discrete analogue of the Itô formula, in its simplest setting, is then stated as follows, which can be interpreted as a discrete version of the well-known Itô formula.

Theorem 1. (Fujita and Kawanishi [5]) (a) *For any* $f : \boldsymbol{Z} \to \boldsymbol{R}$, *we have*

$$f(B_{t+1}) - f(B_t) = \frac{f(B_t + 1) - f(B_t - 1)}{2} Z_{t+1} + \frac{f(B_t + 1) - 2f(B_t) + f(B_t - 1)}{2}.$$

(b) *For any* $f : \boldsymbol{Z} \times \boldsymbol{N} \to \boldsymbol{R}$, *we have*

$$\begin{aligned} &f(B_{t+1}, t+1) - f(B_t, t) \\ &= \frac{f(B_t + 1, t+1) - f(B_t - 1, t+1)}{2} Z_{t+1} \\ &+ \frac{f(B_t + 1, t+1) - 2f(B_t, t+1) + f(B_t - 1, t+1)}{2} \\ &+ f(B_t, t+1) - f(B_t, t). \end{aligned}$$

Proof. We only give the proof of (a). The case of (b) can be proved similarly. Taking account of $B_{t+1} = B_t + 1$ or $B_{t+1} = B_t - 1$ according to $Z_{t+1} = 1$ or $Z_{t+1} = -1$, respectively, we infer that if $Z_{t+1} = 1$,

$$\begin{aligned} f(B_{t+1}) - f(B_t) &= f(B_t + 1) - f(B_t) \\ &= \frac{f(B_t + 1) - f(B_t - 1)}{2} \cdot 1 + \frac{f(B_t + 1) - 2f(B_t) + f(B_t - 1)}{2}, \end{aligned}$$

and if $Z_{t+1} = -1$,

$$\begin{aligned} f(B_{t+1}) - f(B_t) &= f(B_t - 1) - f(B_t) \\ &= \frac{f(B_t + 1) - f(B_t - 1)}{2} \cdot (-1) + \frac{f(B_t + 1) - 2f(B_t) + f(B_t - 1)}{2}, \end{aligned}$$

which is the formula (a).

Now, our basic underlying price process $\{X_t\}_{t=0,1,2\cdots}$ is assumed to be governed by the following discrete stochastic processes.

$$X_{t+1} - X_t = \mu(X_t, t) + \sigma(X_t, t)(B_{t+1} - B_t), \qquad t = 0, 1, 2, \cdots, \quad (3)$$

where μ, σ are given continuous functions. Discrete analogue of the Itô formula for (3) is then expressed as follows.

Proposition 1. (a) *For any* $f : \boldsymbol{R} \to \boldsymbol{R}$, *we have*

$$\begin{aligned} &f(X_{t+1}) - f(X_t) \\ &\quad = \frac{f(X_t + \mu_t + \sigma_t) - f(X_t + \mu_t - \sigma_t)}{2} Z_{t+1} \\ &\quad\quad + f(X_t + \mu_t) - f(X_t) \\ &\quad\quad + \frac{f(X_t + \mu_t + \sigma_t) - 2f(X_t + \mu_t) + f(X_t + \mu_t - \sigma_t)}{2}, \end{aligned} \quad (4)$$

where the use of abbreviations $\mu_t := \mu(X_t, t)$, $\sigma_t := \sigma(X_t, t)$ *is made.*
(b) *For any* $f : \boldsymbol{R} \times \boldsymbol{N} \to \boldsymbol{R}$, *we have*

$$\begin{aligned} &f(X_{t+1}, t+1) - f(X_t, t) \\ &\quad = \frac{f(X_t + \mu_t + \sigma_t, t+1) - f(X_t + \mu_t - \sigma_t, t+1)}{2} Z_{t+1} \\ &\quad\quad + (f(X_t + \mu_t, t+1) - f(X_t, t+1)) \\ &\quad\quad + \frac{f(X_t + \mu_t + \sigma_t, t+1) - 2f(X_t + \mu_t, t+1) + f(X_t + \mu_t - \sigma_t, t+1)}{2} \\ &\quad\quad + (f(X_t, t+1) - f(X_t, t)). \end{aligned} \quad (5)$$

Proof. We just provide a sketch of proof of (a). In view of $Z_{t+1} = B_{t+1} - B_t$, we learn that if $Z_{t+1} = 1$, then it follows that

$$\begin{aligned} &f(X_{t+1}) - f(X_t) \\ &\quad = f(X_t + \mu_t + \sigma_t) - f(X_t) \\ &\quad = \frac{f(X_t + \mu_t + \sigma_t) - f(X_t + \mu_t - \sigma_t)}{2} \cdot 1 \\ &\quad\quad + f(X_t + \mu_t) - f(X_t) \\ &\quad\quad + \frac{f(X_t + \mu_t + \sigma_t) - 2f(X_t + \mu_t) + f(X_t + \mu_t - \sigma_t)}{2}, \end{aligned}$$

while if $Z_{t+1} = -1$, then it follows that

$$
\begin{aligned}
& f(X_{t+1}) - f(X_t) \\
& \quad = f(X_t + \mu_t - \sigma_t) - f(X_t) \\
& \quad = \frac{f(X_t + \mu_t + \sigma_t) - f(X_t + \mu_t - \sigma_t)}{2} \cdot (-1) \\
& \qquad + f(X_t + \mu_t) - f(X_t) \\
& \qquad + \frac{f(X_t + \mu_t + \sigma_t) - 2f(X_t + \mu_t) + f(X_t + \mu_t - \sigma_t)}{2}.
\end{aligned}
$$

Thus we have established (4).

Remark 1. If the time step is taken to be δdt so that (3) turns into

$$X_{t+\delta dt} - X_t = \mu(X_t, t)\delta dt + \sigma(X_t, t)\sqrt{\delta dt}(B_{t+\delta dt} - B_t),$$

we then find, parallel to (4),

$$
\begin{aligned}
& f(X_{t+\delta dt}) - f(X_t) \\
& \quad = \frac{f(X_t + \mu_t \delta dt + \sigma_t \sqrt{\delta dt}) - f(X_t + \mu_t \delta dt - \sigma_t \sqrt{\delta dt})}{2} Z_{t+\delta dt} \\
& \qquad + f(X_t + \mu_t \delta dt) - f(X_t) \\
& \qquad + \frac{f(X_t + \mu_t \delta dt + \sigma_t \sqrt{\delta dt}) - 2f(X_t + \mu_t \delta dt) + f(X_t + \mu_t \delta dt - \sigma_t \sqrt{\delta dt})}{2}.
\end{aligned}
$$

Letting $\delta \to 0$ and invoking the central limit theorem, we see that

$$\frac{Z_{t+\delta t}}{\sqrt{\delta dt}} = \frac{B_{t+\delta t} - B_t}{\sqrt{\delta dt}} \to dB_t$$

in distribution. Consequently, we recover the well known Itô formula

$$df(X_t) = f'(X_t)\sigma_t dB_t + f'(X_t)\mu_t dt + \frac{1}{2} f''(X_t)\sigma_t^2 dt.$$

This is just a heuristic argument, which, however, can be made rigorously under suitable assumptions and justifies our discretization scheme at the same time.

Remark 2. If we take

$$\mu(X_t, t) = \mu_* X_t, \qquad \sigma(X_t, t) = \sigma_* X_t$$

in (3) for positive constants μ_* and σ_*, we obtain a discrete analogue of the well-known Black-Scholes (BS) model [2]. Same scaling as above in Remark 1 leads us to the continuous BS model as well as the famous Black-Scholes formula for European call options. We refer to [4].

3. Optimal portfolio problem

As a first example concerning applications of our discrete analogue of the Itô formula, we analyze the optimal portfolio problem in discrete-time setting. We follow the argument in [7] with enlarged examples.

Suppose a controlled price process $\{X_t\}_{t=0,1,2,\cdots}$ is given by

$$X_{t+1} - X_t = \mu(X_t, t, u_t) + \sigma(X_t, t, u_t)(B_{t+1} - B_t), \qquad t = 0, 1, 2, \cdots, \tag{6}$$

where u_t is measurable with respect to $\sigma(B_k \,|\, k = 0, 1, \cdots, t)$ $(t = 0, 1, 2, \cdots)$, and should be interepreted as the adapted control.

The aim is then to determine the admissible control $\{u_t\}_{t=0,1,2,\cdots}$ which maximizes certain functional involving the utility. To be precise, we wish to solve the problem:

$$V(x, t) := \sup_{\{u_s\}_{s=t}^{T-1}} J(x, t, \{u_s\}_{s=t}^{T-1}), \tag{7}$$

where $T \in N$ and we have defined

$$J(x, t, \{u_s\}_{s=t}^{T-1}) := E^{x,t}\Big[\sum_{k=t}^{T-1} U_1(X_k, k, u_k) + U_2(X_T, T) \,\Big|\, X_t = x\Big].$$

Here U_1, U_2 are utility functions, which are increasing and strictly concave in X_k.

The next two theorems are established in [7]. In the below, for convenience of notation, we put the right hand side of (5) as follows:

$$\begin{aligned}\mathcal{L}_X f(X_t, t) :=\ & f(X_t + \mu_t, t+1) - f(X_t, t+1) \\ & + \frac{f(X_t + \mu_t + \sigma_t, t+1) - 2f(X_t + \mu_t, t+1) + f(X_t + \mu_t - \sigma_t, t+1)}{2} \\ & + f(X_t, t+1) - f(X_t, t),\end{aligned} \tag{8}$$

where we have written for short

$$\mu_t := \mu(X_t, t, u_t), \qquad \sigma_t := \sigma(X_t, t, u_t).$$

The discrete Hamilton-Jacobi-Bellman (d-HJB) equation derived in [7] is now stated as follows, which features the property of the value function $V(x, t)$ and hence gives a solution to the stochastic control problem (7).

Theorem 2. (Discrete Hamilton-Jacobi-Bellman (d-HJB) equation) *For $t = 0, 1, \cdots, T$, we have*

$$\begin{aligned} \sup_{\{u_s\}_{s=t}^{T-1}} \{\mathcal{L}_X^u V(x,t) + U_1(x,t,u_t)\} &= 0 \\ V(x,T) &= U_2(x,T), \end{aligned} \tag{9}$$

where $\mathcal{L}_X$ is defined in (8).

Proof. The so-called Bellman principle is applied to obtain

$$V(x,t) = \sup_{\{u_s\}_{s=t}^{T-1}} E^{x,t}\Big[U_1(x,t,u_t) + V(X_{t+1}, t+1)\Big]$$
$$V(x,T) = U_2(x,T).$$

We now appeal to Proposition 1 (b) to conclude the demonstration of the theorem.

Theorem 3. (Verification theorem) *Let $W(x,t)$ solves the discrete Hamilton-Jacobi-Bellman equation (9):*

$$\sup_{\{u_s\}_{s=t}^{T-1}} \{\mathcal{L}_X^u W(x,t) + U_1(x,t,u_t)\} = 0,$$
$$W(x,T) = U_2(x,T).$$

Then we have

$$W(x,t) \geq J(x,t,\{u_s\}_{s=t}^{T-1}), \tag{10}$$

for every $x \in \boldsymbol{R}$, $t = 0, 1, 2, \cdots, T-1$ and adapted $\{u_t\}$. Furthermore, if for every $x \in \boldsymbol{R}$, $t = 0, 1, 2, \cdots, T-1$, then there exists an adapted $\{u_k^*\}$ with

$$u_k^* \in \arg \sup_{\{u_s\}_{s=t}^{T-1}} (\mathcal{L}_X^u W(k, X_k^*) + U_1(x, X_k^*, u_k)),$$

for every $t \leq k \leq T-1$, where X_k^* is the controlled process corresponding to u_k^* through (6), then we obtain

$$W(x,t) = V(x,t) = J(x,t,\{u_s\}_{s=t}^{T-1}).$$

Proof. It suffices to show that

$$W(x,t) \geq E^{x,t}\Big[\sum_{k=t}^{T} U_1(X_k, k, u_k) + U_2(X_T, T) \,\Big|\, X_t = x\Big], \tag{11}$$

for every adapted $\{u_t\}$. Since W is a solution of the d-HJB equation, we see that

$$\mathcal{L}_X^u W(x,t) + U_1(x,t,u_t) \le 0. \tag{12}$$

We then apply Theorem 2 to find

$$W(X_T,T) - W(X_t,t) = \sum_{k=t}^{T-1} \mathcal{L}_X^u W(X_k,k) +$$
$$+ \sum_{k=t}^{T-1} \frac{W(X_k+\mu_k+\sigma_k,t+1) - W(X_k+\mu_k-\sigma_k,t+1)}{2} Z_{k+1}.$$

Taking expectations and employing the inequality (12), we obtain the inequality (11) and the proof is completed.

Concerning the theory of optimal portfolio problem in continuous time, we refer for instance to [1][9].

Now we present examples of the d-HJB equation to illustrate this theory.

Example 1. We assume in (6) that $\mu(X,t,u) = \mu u X$ and $\sigma(X,t,u) = \sigma u X$ with $\mu \ge 0$ and $\sigma > \mu$ being constants. As to utility functions, we first take $U_1 \equiv 0$ and $U_2 = \sqrt{x}$. Therefore, the d-HJB equation (9) becomes

$$\begin{aligned} \sup_{\{u_s\}_{s=t}^{T-1}} \Big\{ & V((1+\mu u_t)x, t+1) - V(x,t+1) \\ & + \frac{V((1+(\sigma+\mu)u_t)x,t+1) - 2V(x,t+1) + V((1-(\sigma-\mu)u_t)x,t+1)}{2} \\ & + V(x,t+1) - V(x,t) \Big\} = 0 \end{aligned} \tag{13}$$
$$V(x,T) = \sqrt{x}.$$

We will seek a solution of the form

$$V(x,t) = g(t)\sqrt{x} \tag{14}$$

where $g(T) = 1$. Inserting (14) into (13) we deduce that

$$\begin{aligned} \sup_{0 \le u_t \le (\sigma-\mu)^{-1}} \Big\{ & g(t+1) \cdot \\ & \cdot \frac{2(\sqrt{(1+\mu u_t)x} - \sqrt{x}) + \sqrt{(1+(\sigma+\mu)u_t)x} + \sqrt{(1-(\sigma-\mu)u_t)x}}{2} \\ & - g(t)\sqrt{x} \Big\} = 0. \end{aligned}$$

The maximization is attained by some constant strategy. For instance, if $\mu \equiv 0$, then the optimal strategy is given by $u_t \equiv 0$ and hence we obtain $g(t) \equiv 1$ as well as $V(x,t) \equiv \sqrt{x}$.

As the second possibility of utility functions, we take $U_1 \equiv 0$ and $U_2 = \log x$; we seek a solution of the form $V(x,t) = \log x + g(t)$ with $g(T) = 0$. Putting this form into (13), we compute

$$\sup_{0 \le u_t \le (\sigma-\mu)^{-1}} \Big\{ \log(1+\mu u_t)(1+(\sigma+\mu)u_t)^{1/2}(1-(\sigma-\mu)u_t)^{1/2} + g(t+1) - g(t) \Big\} = 0.$$

The optimal strategy is again offered by some constant u_t. For instance, if $\mu \equiv 0$, then the optimal strategy is given by $u_t \equiv 0$ and hence we obtain $g(t) \equiv 0$ as well as $V(x,t) \equiv \log x$.

Example 2. We assume in (6) that $\mu(X,t,u) = u$ and $\sigma(X,t,u) = \sigma u$ with $\sigma > 1$. As to utility functions, we first take $U_1 \equiv 0$ and $U_2 = \sqrt{x}$ as before. Therefore, the d-HJB equation (9) becomes

$$\begin{aligned} &\sup_{\{u_s\}_{s=t}^{T-1}} \Big\{ V(x+u_t,t+1) - V(x,t+1) + \\ &\quad \frac{V(x+(1+\sigma)u_t,t+1) - 2V(x+u_t,t+1) + V(x+(1-\sigma)u_t,t+1)}{2} \\ &\quad + V(x,t+1) - V(x,t) \Big\} = 0 \\ &V(x,T) = \sqrt{x}. \end{aligned} \tag{15}$$

We will seek a solution of the form

$$V(x,t) = g(t)\sqrt{x} \tag{16}$$

where $g(T) = 1$. Inserting (16) into (15) we infer that

$$\sup_{\{u_t\}\in\mathcal{A}} \Big\{ g(t+1)\frac{\sqrt{x+(1+\sigma)u_t} + \sqrt{x-(\sigma-1)u_t}}{2} - g(t)\sqrt{x} \Big\} = 0.$$

The maximization is attained by the optimal strategy

$$u_t = \frac{2}{\sigma^2-1}x. \tag{17}$$

Plugging (17) back into (15) we obtain

$$g(t) = \Big(\frac{1}{2}\Big)^{T-t}\Big(\sqrt{\frac{\sigma+1}{\sigma-1}} + \sqrt{\frac{\sigma-1}{\sigma+1}}\Big)^{T-t}$$

and the corresponding $V(x,t)$.

As the second possibility of utility functions, we take $U_1 \equiv 0$ and $U_2 = -e^{-ax}$ with $a > 0$; we seek a solution of the form $V(x,t) = -g(t)e^{-ax}$ with $g(T) = 1$. Inserting this form into (15), we find that

$$\sup_{u_t} \left\{ -g(t+1)\frac{e^{-a(x+(1+\sigma)u_t)} + e^{-a(x-(\sigma-1)u_t)}}{2} + g(t)e^{-ax} \right\} = 0.$$

The optimal strategy is given by the constant strategy $u_t = (2a\sigma)^{-1}\log(\sigma+1)/(\sigma-1)$ and thus

$$g(t) = \cosh^{T-t}\left(\frac{1}{2\sigma}\sqrt{\frac{\sigma+1}{\sigma-1}}\right).$$

4. Pricing of Margrabe option

Margrabe option [10] is also called as an exchange option, which gives the holder the right to exchange one X_t^1 share for another X_t^2 share at the maturity T. The payoff value of this option V is thus

$$\max\{X_T^2 - X_T^1, 0\} =: V(X_T^1, X_T^2, T).$$

The theme of our study in this section is to estimate the price of $V(X_t^1, V_t^2, t)$, especially for $t = 0$.

For this purpose, we need to clarify the price process (3) of X_t^1 and X_t^2, respectively. It then turns out that, for convenience's sake, we had better introduce a new stochastic variable

$$Y_t := \frac{X_t^2}{X_t^1},$$

which means that the random variable X_t^1 is set to be numeraire.

Now the stochastic process of X_t^1 is assumed to be given by

$$X_{t+1}^1 - X_t^1 = aX_t^1 + bX_t^1 Z_{t+1}^1,$$

where a, b are positive constants, and $\{Z_t^1\}_{t=0,1,2,\cdots}$ $(Z_0^1 = 0)$ fulfill (2). If we define $\alpha, \beta \in \boldsymbol{R}$ by

$$a = e^{\alpha}\cosh\beta, \qquad b = e^{\alpha}\sinh\beta,$$

then, taking account of the discrete analogue of the Itô formula, we deduce that

$$X_t^1 = X_0^1 e^{\alpha t + \beta B_t^1}, \quad \text{where} \quad B_t^1 = \sum_{n=1}^{t} Z_n^1.$$

For simplicity, we additionally assume $\alpha = 0$, since this term corresponds to the drift, and we may discard it without altering the essential features. We also put $X_0^1 = 1$.

Along a similar way, we realize that the price process of X_t^2 in our model should be given by

$$X_t^2 := X_t^1 e^{\gamma B_t^2}, \quad \text{where} \quad B_t^2 = \sum_{n=1}^{t} Z_n^2.$$

Here $\{Z_t^2\}_{t=0,1,2,\cdots}$ $(Z_0^2 = 0)$ denotes another i.i.d. random variables, which is assumed to be independent of $\{Z_t^1\}$.

To summarize, we deal with the next price processes.

$$\begin{aligned} X_t^1 &= e^{\beta B_t^1}, \quad \text{where} \quad B_t^1 = \sum_{n=1}^{t} Z_n^1, \ B_0^1 = 0, \quad \text{and } \beta > 0 \\ X_t^2 &= X_t^1 e^{\gamma B_t^2}, \quad \text{where} \quad B_t^2 = \sum_{n=1}^{t} Z_n^2, \ B_0^1 = 0, \quad \text{and } \gamma > 0. \end{aligned} \tag{18}$$

For slightly different models, we refer to [8].

Our task is then to evaluate

$$\begin{aligned} V_0 &:= E^Q[V(X_T^1, X_T^2, T)] = E^Q[\max\{X_T^2 - X_T^1, 0\}] \\ &= E^Q[X_T^1 \max\{Y_T - 1, 0\}] = X_0^1 E^Q[\max\{Y_T - 1, 0\}], \end{aligned}$$

where Q represents the risk-neutral probability measure. We prove the next theorem.

Theorem 4. *Under the price processes (18), the present value V_0 of the Margrabe option V, whose payoff fnction is given by $V(X_T^1, X_T^2, T) = \max\{X_T^2 - X_T^1, 0\}$, is expressed as*

$$V_0 = \sum_{i=k_0+1}^{T} \binom{T}{i} \left(\frac{1 - e^{-\gamma}}{e^{\gamma} - e^{-\gamma}}\right)^i \left(\frac{e^{\gamma} - 1}{e^{\gamma} - e^{-\gamma}}\right)^{T-i} e^{(2i-T)\gamma} - 1,$$

where k_0 is the integer such that $(T-2)/2 < k_0 \leq T/2$.

Proof. Let $P(Z_t^2 = 1) =: p$, $P(Z_t^2 = -1) = 1 - p$ under the measure Q, and define $N_t := \#\{i = 1, 2, \cdots, t \mid Z_i^2 = 1\}$. We learn that

$$P(N_t = k) = \binom{T}{k} p^k (1-p)^{T-k}, \qquad k = 0, 1, \cdots, T.$$

Taking into account that $Y_T = X_T^2/X_T^1 > 1$ is equivalent to $e^{(2k-T)\gamma} > 1$, we define the integer k_0 such that

$$\frac{T-2}{2} < k_0 \leq \frac{T}{2}.$$

We then observe that

$$V_0 = \sum_{i=k_0+1}^{T} \binom{T}{i} p^i (1-p)^{T-i} e^{(2i-T)\gamma} - 1. \tag{19}$$

Next we determine the risk-neutral probability measure p, which is characterized by the property that $\{Y_t\}_{t=0,1,\cdots,T}$ turns out to be martingale. It follows that $E^Q[e^{\gamma Z_{t+1}^2}] = 1$ and therefore

$$p = \frac{1-e^{-\gamma}}{e^{\gamma} - e^{-\gamma}}. \tag{20}$$

Placing (20) into (19), we arrive at the desired conclusion.

5. Discussions

We have developed the applications of the discrete analogue of the Itô formula, which is recently introduced by one of the authors. In the discrete random world, the formula provides a fundamental tool of analysis and corresponds to the glorious Itô formula in continuous time. In terms of this analogous formula, the bridge between the discrete and the continuous processes may be well understood.

To check the effectiveness of our analogous formula, we here discuss the problem of optimal portfolio decision and the pricing of exchange options. The discrete analogue of the Itô formula is seen to work as a central player. Moreover, both examples result in the combinatorial characterization and thus we believe that our discrete analogue of the Itô formula naturally implies the accessibility of numerical implementation.

The application of the discrete analogue of the Itô formula is not limited by the examples pursued here. It would be interesting to revisit several issues in discrete stochastic processes within the context of the current formula. In particular, we are now investigating some topics from the insurance mathematics [6], which will be reported in the near future.

References

1. Björk, T.: Arbitrage Theory in Continuous Time, 2nd ed. Oxford Univ. Press, Oxford (2004)
2. Black, F., Scholes, M.: The pricing of options and corporate liabilities. J. Political Economy **81**, 637–659 (1973)
3. Fujita, T.: Introduction to the Stochastic Analysis for Financial Derivatives (Finance no Kakuritsu-Kaiseki Nyumon) (in Japanese). Kodanshya, Tokyo (2002)
4. Fujita, T.: Random Walks and the Stochastic Analysis (Random Walk to Kakuritsu-Kaiseki) (in Japanese). Nihon Hyoron-shya, Tokyo (2008)
5. Fujita, T., Kawanishi, Y.: A proof of Itô's formula using a discrete Itô's formula. Stud. Scienti. Math. Hungarica **45**, 125–134 (2008)
6. Gerber, H.U., Shiu, E.S.W.: Actuarial bridges to dynamic hedging and option pricing. Insurance: Mathematics and Economics **18**, 183–218 (1996)
7. Ishimura, N., Mita, Y.: A note on the optimal portfolio problem in discrete processes. Kybernetika **45**, 681–688 (2009)
8. Kawai, N.: Risan Kakuritsu Kaiseki deno Sokudo Henkan ni tsuite – Risan Margrabe option no pricing (On the measure transformation in the discrete stochastic analysis – The pricing of discrete Margrabe option) (in Japanese). Thesis for the Master-course degree, Graduate School of Economics, Hitotsubashi University (2009)
9. Korn, R., Korn, E.: Option Pricing and Portfolio Optimization. Graduate Studies in Mathematics 31, American Mathematical Society, Rhode Island (2001)
10. Margrabe, W.: The value of an option to exchange one asset for another. J. Finance **33**, 177–186 (1978)

Subject Index

S. Kusuoka and T. Maruyama (eds.), *Advances in Mathematical Economics Volume 16*, DOI: 10.1007/978-4-431-54114-1,

Advances in
MATHEMATICAL
ECONOMICS

Instructions for Authors

A. General

1. Papers submitted for publication will be considered only if they have not been and will not be published elsewhere without permission from the publisher and the Research Center for Mathematical Economics.

2. Every submitted paper will be subject to review. The names of reviewers will not be disclosed to the authors or to anybody not involved in the editorial process.

3. The authors are asked to transfer the copyright to their articles to Springer if and when these are accepted for publication.

The copyright covers the exclusive and unlimited rights to reproduce and distribute the article in any form of reproduction. It also covers translation rights for all languages and countries.

4. Manuscript must be written in English. Its pdf file should be submitted by e-mail: maruyama@econ.keio.ac.jp

Office of Advances in
Mathematical Economics
c/o Professor Toru Maruyama
Department of Economics
Keio University
2-15-45, Mita Minato-ku,
Tokyo 108-8345, JAPAN

5. Twenty-five (25) **offprints** of each paper will be supplied free of charge. Additional offprints can be ordered at cost price.

B. Preparation of Manuscript

1. Manuscripts should be submitted in the pdf format. If this is not possible, two printouts of the manuscript must be submitted to the above postal address.

Manuscripts should be written in **LaTeX**. Please use Springer's LaTeX macro package (download from `ftp://ftp.springer.de/pub/tex/latex/svjour/global/`). After acceptance, sending the original source (including all style files and figures) and a pdf (compiled output) are required.

Authors wishing to include figures, tables, or text passages that have already been published elsewhere are required to **obtain permission** from the copyright owner(s) for both the print and online format.

2. **The title page** should include:
- The name(s) of the author(s)
- A concise and informative title
- The affiliation(s) and address(es) of the author(s)
- The e-mail address, telephone and fax numbers of the corresponding author

Please provide an **abstract** less than 100 words. The abstract should not contain any undefined abbreviations or unspecified references.

Please provide 4 to 6 **keywords** which can be used for indexing purposes.

3. Please use the decimal system of **headings** with no more than three levels.

Abbreviations should be defined at first mention and used consistently thereafter.

Footnotes can be used to give additional information, which may include the citation of a reference included in the reference list. They should not consist solely of a reference citation, and they should never include the bibliographic details of a reference. They should also not contain any figures or tables. Footnotes to the text are numbered consecutively; those to tables should be indicated by superscript lower-case letters (or asterisks for significance values and other statistical data). Footnotes to the title or the authors of the article are not given reference symbols. Always use footnotes instead of endnotes.

4. **The Journal of Economic Literature index number (JEL classification)** should be indicated and the statement of the **2010 Mathematics Subject Classification (MSC) numbers** is desirable. You can check JEL classification with Internet at `http://ideas.repec.org/JEL/` as well as 2010 MSC numbers at `http://www.ams.org/msc`.

5. **Main text**: All **tables and figures** must be cited in the text and numbered consecutively with Arabic numerals according to the sequence in which they are cited.

For each table, please supply a table caption (title) explaining the components of the table. Identify any previously published material by giving the original source in the form of a reference at the end of the table caption. When preparing your tables and figures, size them to fit in the column width.

Short **equations** can be run in with the text. Equations that are displayed on a separate line should be numbered.

6. **Reference citations** in the text should be identified by numbers in square brackets. Some examples:
1. Negotiation research spans many disciplines [3].
2. This result was later contradicted by Becker and Seligman [5].
3. This effect has been widely studied [1–3, 7].

The list of references should only include works that are cited in the text and that have been published or accepted for publication. Personal communications and unpublished works should only be mentioned in the text. Do not use footnotes or endnotes as a substitute for a reference list. The entries in the list should be numbered consecutively.

•*Journal article*
Hamburger, C.: Quasimonotonicity, regularity and duality for nonlinear systems of partial differential equations. Ann. Mat. Pura. Appl. **169**, 321–354 (1995)
•*Article by DOI*
Sajti, C.L., Georgio, S., Khodorkovsky, V., Marine, W.: New nanohybrid materials for biophotonics. Appl. Phys. A (2007). doi:10.1007/s00339-007-4137-z
•*Book*
Geddes, K.O., Czapor, S.R., Labahn, G.: Algorithms for Computer Algebra. Kluwer, Boston (1992)
•*Book chapter*
Broy, M.: Software engineering - from auxiliary to key technologies. In: Broy, M., Denert, E. (eds.) Software Pioneers, pp. 10–13. Springer, Heidelberg (2002)
•*Online document*
Cartwright, J.: Big stars have weather too. IOP Publishing PhysicsWeb. `http://physicsweb.org/articles/news/11/6/16/1` (2007). Accessed 26 June 2007

Please use the standard abbreviation of a journal's name according to the ISSN List of Title Word Abbreviations, see `http://www.issn.org/2-22660-LTWA.php`

7. The purpose of the **author proof** is to check for typesetting or conversion errors and the completeness and accuracy of the text, tables and figures. Substantial changes in content, e.g., new results, corrected values, title and authorship, are not allowed without the approval of the Editor. After online publication, further changes can only be made in the form of an Erratum, which will be hyperlinked to the article.

8. Please use the standard **mathematical notation** for formulae, symbols etc.: Italic for single letters that denote mathematical constants, variables, and unknown quantities Roman/upright for numerals, operators, and punctuation, and commonly defined functions or abbreviations, e.g., cos, det, e or exp, lim, log, max, min, sin, tan, d (for derivative) Bold for vectors, tensors, and matrices.

The manufacturer's authorised representative in the EU is Springer Nature Customer Service Centre GmbH, Europaplatz 3, 69115 Heidelberg, Germany. If you have any concerns regarding our products, please contact ProductSafety@springernature.com

Printed and bound by CPI Group (UK) Ltd, Croydon, CR0 4YY
25/09/2026
02222894-0013